TECHNOLOGY AND ENTREPÔT COLONIALISM IN SINGAPORE, 1819-1940

The **Institute of Southeast Asian Studies (ISEAS)** was established as an autonomous organization in 1968. It is a regional centre dedicated to the study of socio-political, security and economic trends and developments in Southeast Asia and its wider geostrategic and economic environment. The Institute's research programmes are the Regional Economic Studies (RES, including ASEAN and APEC), Regional Strategic and Political Studies (RSPS), and Regional Social and Cultural Studies (RSCS).

ISEAS Publishing, an established academic press, has issued more than 2,000 books and journals. It is the largest scholarly publisher of research about Southeast Asia from within the region. ISEAS Publishing works with many other academic and trade publishers and distributors to disseminate important research and analyses from and about Southeast Asia to the rest of the world.

TECHNOLOGY AND ENTREPÔT COLONIALISM IN SINGAPORE, 1819–1940

GOH CHOR BOON

INSTITUTE OF SOUTHEAST ASIAN STUDIES
Singapore

First published in Singapore in 2013 by
ISEAS Publishing
Institute of Southeast Asian Studies
30 Heng Mui Keng Terrace
Pasir Panjang
Singapore 119614

E-mail: publish@iseas.edu.sg
Website: <http://bookshop.iseas.edu.sg>

All rights reserved. No part of this publication may be reproduced, stored in a retrieval system, or transmitted in any form or by any means, electronic, mechanical, photocopying, recording or otherwise, without the prior permission of the Institute of Southeast Asian Studies.

© 2013 Institute of Southeast Asian Studies, Singapore

The responsibility for facts and opinions in this publication rests exclusively with the author and his interpretations do not necessarily reflect the views or the policy of the publishers or their supporters.

ISEAS Library Cataloguing-in-Publication Data

Goh, Chor Boon.
    Technology and entrepôt colonialism in Singapore, 1819–1940.
    1. Singapore—Economic conditions.
    2. Technology—Social aspects—Singapore—History.
    3. Entrepôt trade—Singapore—History.
    I. Title.
    II. Series: Modern economic history of Southeast Asia
HC445.8 G614   2013

ISBN 978-981-4414-08-1 (soft cover)
ISBN 978-981-4414-09-8 (e-book, PDF)

Cover photo: Cavenagh Bridge, 1880 to 1892.
Photo credit: Tropenmuseum, Amsterdam, Coll. no. 60000046.

Typeset by Superskill Graphics Pte Ltd
Printed in Singapore by Markono Print Media Pte Ltd.

Contents

List of Figures and Tables — vi
Notes on Currency — vi
Introduction — 1
1 Technology and the British Empire — 13
2 Pioneers of Change: Entrepreneurs and Engineers — 35
3 Maritime Technology and Development of the Port — 64
4 Introducing Technological Systems — 93
5 Sanitation and Public Health — 120
6 Agriculture and Colonial Science — 143
7 Food and Singapore Cold Storage — 168
8 Politics of Imperial Education — 196
9 Technology Transfer and Limited Industrial Growth — 221
Conclusion — 240
Bibliography — 247
Index — 265

List of Figures and Tables

Figure 7.1	"Look Before You Leap!", *Straits Times*, 21 November 1936	185
Figure 7.2	"Why doesn't she go to the Cold Storage?", *Straits Times*, 18 November 1936	186
Table 4.1	Government Telegraphy Operations	98
Table 8.1	Expenditure on Education in the Federated Malay States, 1875–1900	200
Table 8.2	Return of Industrial Scholarships and Apprenticeships	204
Table 8.3	Occupations of Candidates applying for Industrial Scholarships	205
Table 8.4	Number of Professionals in the Straits Settlements in 1925	206
Table 8.5	Attendance in Science-based Subjects, 1922–23	207
Table 9.1	Chinese Manufacturing Activity in 1928	223

Notes on Currency

Unless otherwise stated, the unit of currency used throughout the text is the Straits dollar. Prior to 1906, the Straits dollar fluctuated between 4 shillings and 6 pence in 1874 and 1 shilling and 8½ pence in 1902. A new Straits dollar was introduced in 1903 and the rate of exchange fixed at 2 shillings 4 pence in 1906 and remained at this level up to 1967.

Introduction

When the city-state of Singapore gained its independence in 1965, there was the question of whether the government should bring down the statue of Sir Stamford Raffles, the "founder" of Singapore and standing proudly in front of the iconic Victoria Memorial Hall. The statue was unveiled when the colony celebrated the fiftieth anniversary of the reign of Queen Victoria in 1887.[1] In the words of Lee Kuan Yew:

> Investors wanted to see what a new socialist government in Singapore was going to do to the statue of Raffles. Letting it remain would be a symbol of public acceptance of the British heritage and could have a positive effect. I had not looked at in that way, but was quite happy to leave this monument because he was the founder of modern Singapore. If Raffles had not come here in 1819 to establish a trading post, my great-grandfather would not have migrated to Singapore from Dapu county in Guangdong province, southeast China. The British created an emporium that offered him, and many thousands like him, the opportunity to make a better living than in their homeland which was going through turmoil and chaos as the Qing dynasty declined and disintegrated.[2]

For many former colonies of the British and other European empires, the memory of the colonial period is often a painful one, a national humiliation of conquest, military occupation and subservience. Independence was more often than not greeted as a celebration of liberation. For the island-state of Singapore, British colonialism and its ideology of "civilizing mission" did provide the thousands of Chinese, Indians, Malays and Europeans "the opportunity to make a better living". Together with the British and other European entrepreneurs, engineers, missionaries, and municipal administrators, these forefathers laid a strong foundation of a modern city. The story of their effort started one morning in the month of January of the year 1819.

On 29 January 1819, Thomas Raffles landed on the white sandy beach of Singapore, near the mouth of the Singapore River. According to an eyewitness, Raffles was accompanied by two white men and a sepoy who carried a musket.[3] Anchored out at sea and ready for action was the British gunboat *Indiana*. In one of his letters to his friends in London, Raffles wrote: "At Singapore I found advantages far superior to what Rhio afforded … our station completely outflanks the Straits of Malacca, and secures a passage for our China ships at all times, and under all circumstances. It has further been my good fortune to discover one of the most safe and extensive harbours in these seas, with every facility for protecting shipping in time of war … [Singapore] will soon rise to importance…".[4] The island was ceded to Britain, represented by the employees of the East India Company, with a stroke of the quill pen — and not the gun.[5] The "gunboat diplomacy" succeeded without actually using force, though all parties knew it was available in case of need. In all probability, it was not the first time the Malay villagers and their headman saw the firepower-mounted British steamships that would soon force its way up the Irrawaddy River in 1824 and, a decade or so later, the world's first all-iron steamer, the *Nemesis*, prised open the "Middle Kingdom" of China to European traders and missionaries in 1841. Since the seventeenth century, the Portuguese and the Dutch were marauding with their gunboats in the Indonesian archipelago with the economic objective of controlling the spice trade. But from the middle of the eighteenth century, the Union Jack of Great Britain led the scramble for tropical colonial empires — made easy and swift by a few technical innovations. One new institutional entity for exerting British power and influence, and backed by the Royal Navy, was the establishment of government-run trade entrepôts. They could serve as a naval base, a point of safety for warehousing and distributing the new output of the rapidly industrializing Western world and for bulking raw materials for European industry. As part of Britain's overseas empire, Singapore's ascendancy to supremacy over other trading ports in the regional port hierarchies was due to the functions the port offered on imperial sea routes in the region. However, one can argue that Singapore — known as Temasek some five hundred years ago — was already a thriving seaport in the fourteenth century, frequented by Javanese and Chinese traders and used a hideout

Introduction 3

by pirates.⁶ But it was also exposed to the internecine squabbles between the rival expanding empires of Majapahit and Thailand. The unique geographical location of the island was not of importance in ancient days and, devoid of natural resources, Singapore remained as a small trading village inhabited mainly by boat-dwellers. What gave a new twist to the traditional pattern, in Singapore's case, was that nineteenth century Singapore was the product of the forces of Britain's Industrial Revolution. Technological change and the rapid industrialization of Western nations contributed to the creation of colonial empires, especially in Asia and Africa, and stimulated the demand for the products of the tropics. Steamships, railways and telegraphs allowed Europeans to rule their newly acquired colonies efficiently. New industries required new materials from the tropics — cotton and indigo, palm oil, copper, gutta-percha, tin and rubber. An increasingly affluent Western consumer society demanded sugar, tea, coffee, cocoa, spices and other tropical goods. There was also massive transfer of technology from the West to Asia and Africa and this, in turn, led to the growth in tropical production of goods destined for the industrial cities in the West. Colonized people were introduced to a whole array of technologies and manufactured goods.

From the 1850s, Britain led the West in the creation of an international trading infrastructure. In the words of Peter Stearns, "[f]orming part of this structure were international trading companies and shipping lines, which were expanded by the technology of the steamship … the globe was shrinking because of industrial technology and new levels of world trade…".⁷ Singapore was like a fresh frontier of the British economic interest in the East, eagerly waiting to receive the benefits of Western scientific and technological advancement and the economics of free trade. While the study of the relationship between technology transfer, modernization, economic development and colonial politics in India has been widely researched — and rightly so since the continent was the strategic centre of the British Empire and the envy of all other Western colonial powers — little has been done on this relationship in the context of an important colonial port-city in the "Far East" of the British Empire. This book hopes to fill this void by examining the role of imported technology on the growth and development of the port-city of Singapore from 1819 to 1940 and set as a case study to illustrate Britain's "civilizing mission". It does not purport to any extensive interpretation of existing or new primary sources relating to science and technology in colonial Singapore. This is because archival reference to science and technology *per se* is piecemeal and incidental in nature and largely drawn from various official reports and personal accounts of contemporary residents and travellers. There is hardly any secondary works done on the role of technology in the growth of the port-city during the period of British rule. It is acknowledged that this study is heavily biased towards Western samples mainly because there is no attempt made to look into pertinent Chinese sources that could provide a useful angle of interpretation. However, it could be inferred that such sources are also far and few.

The book serves three purposes. First it hopes to add a new dimension to the historiography of Singapore from a "science, technology and society" perspective. As it stands, the scholarship leans largely on the development of the city-state after 1965 (the year of its independence), particularly in the social, economic and political arena of nation-building, popularly encapsulated in the term "The Singapore Story". How has everyday life changed between 1819 and 1940? How much of this change is due to the introduction of Western innovations in science and technology into the port-city? The important roles of technology and science in transforming the life of the inhabitants in colonial Singapore before 1940 have yet to be sufficiently recognized and documented. This book opens up a new angle in looking at Singapore's development as a port-city, and adds an important dimension in understanding the impact of colonialism in Singapore. Such studies on the impact of science and technology have already been undertaken in places like India, China and Japan, and this book will add to the growing literature in the field.

By providing an overview of technological change in colonial Singapore during the years 1819 to 1940, this book also introduces a historical perspective to the understanding of Singapore's current quest for technological and scientific excellence. While it is true that the antecedents of Singapore's colonial past tend to diminish over time, it is also plausible that explanation for the issues and challenges of Singapore's science and technology policies in today's context can be traced to its historical developments. Much has been written on the technological success of Japan, South Korea, and Taiwan — how they transformed from imitators to innovators — and the process of which has been explained with references to their national histories, in particular, to their encounters with Western science and technology and their subsequent enactment of supportive state policies to develop the indigenous technological base.[8]

Introduction 5

Indeed, a more matching comparison is Hong Kong, an entrepôt under the British rule until 1997. While the remarks on Hong Kong are made in passing, they are nevertheless useful to highlight certain arguments made in the book.

On a broader canvas, the book provides a view of the British Empire from the periphery — by documenting a more integrative interpretation of Singapore's colonial past with that of the Imperial metropolis. Some observers have lamented that in Britain itself, "by the 1980s, colonial history was largely dropped as a subject in Britain's increasingly multi-ethnic schools" and that "citizens in the old colonies" are generally apathetic towards their historical ties.[9] While it is true that there is only limited mention of the British Empire and the period of British rule in Singapore (particularly before 1940) in the school history textbooks, there were public exhibitions and talks organized to prick the collective memory of Singaporeans of the city-state's imperial past. This study hopes to generate readership interest and perhaps open up discussions on the broader question of the relationship between technology and the Singapore society in the period of European imperialism and colonialism. However, this study is not preoccupied with the constitutional, political and economic aspects of British rule in the colony. There are notable works done by local nationalist historians under the so-called "old" or "traditional" imperial history approach who tend to see their work as offering lessons in citizenship and moral instruction relating to issues such as racism, chauvinism and colonialism.[10] This book hopes to encourage local scholars to return to the history of British imperialism in a search for the historical roots of globalization, or to write "new imperial history" as a form of global history. It provides an interdisciplinary, thematic platform to study the representations of empire — issues of imperial science and technology, culture, gender, race, education — which played so central a part in the creation of what many "post-colonist" historians regard as the significant experiences of imperial dominance and imperial subservience.[11]

No study of science, technology and society, however, could hope to cover every aspect of a society's modern social and economic development. In writing this book, I have resisted repeated temptations to follow up with all sorts of intriguing anecdotes and reports that relate to how science and technology had affected everyday life in Singapore during the decades before 1940. Such an embracing subject also demands that the text swings between social history of technology and economic history, and between culture and politics. What I have tried to do is to cover issues in specific areas of technological change which help to illustrate the relationship between society and technology in modern Singapore. Examining areas, such as shipping, port development, telegraphs and wireless, urban water supply and sewage disposal, economic botany, electrification, food production and retailing, science and technical education, and health, this book documents the role of technology and, to a smaller extent, science, in the transformation of colonial Singapore. How did imported technology contribute to the development of the colony? Who were the main agents of change in this process? Was there extensive transfer and diffusion of Western science and technology into the port-city? How did the people respond to change? What is the impact of British colonialism on the development of a technological culture in Singapore? One key argument of the book is that various Western-imported technologies, especially when combined, enhanced the colonial administration's and the local business community's ability to govern the port-city and to expand the entrepôt trading system respectively. While the transfer of these imported technologies gave imperial agents more scope for intervention, the indigenous community gained access to them as well and benefited from their transfer.[12] The central theme that runs through the chapters is that the construction and development of British imperial ports like Singapore through the introduction of Western technologies and institutional practices is carried within the framework of British colonialism — to serve the needs of London and the preservation of British imperial greatness in the Empire.

Chapter One provides an overview on the role of technology in the expansion and development of the British empire. While the power of her technological prowess enabled Great Britain to acquire colonies, British colonial administrators, missionaries and businessmen came with their railroads, steamships, telegraphy and industrial machines — as powerful civilizing artefacts. British officials who planned and administered colonial cities like Calcutta and Singapore had to do so within the overall framework of colonialism. The paternalistic mission stressed on the maintenance of law and order, rather than the progressive uplift of the colonized. The preservation of British power and prestige was all-important. Technology and its transfer served ideological as well as physical purposes. Singapore has no natural resources. What made the fishing village into the vibrant city-state today are the people in its history. Rightly then, Chapter Two examines the agents of change — entrepreneurs, surveyors, architects and engineers — who arrived as part of the diaspora of European skilled immigrants to Singapore and Malaya during the nineteenth century and early twentieth century. They were the harbingers of the "civilizing mission". However, the entrepreneurial role of the Chinese middlemen or compradors — supported by the thousands of immigrants from South China — also contributed significantly to the successful trading networks of the European agency houses. Collectively, people from all over the world came to tap on the wealth generated by the small island as it grew rapidly into one of Britain's premier ports in the East. Chapter Three centres on the development of the Singapore port, from its existence at the Singapore River to the "New Harbour" at the southwestern part of the island. It was a continuing process of expansion and adaptation in response to new maritime technology. To accommodate the increasing size and number of ships and the growing volume of world trade, particularly with the opening of the Suez Canal in 1869, Singapore, together with a whole series of new colonial ports — Port Said, Aden, Karachi, Dakar, and Hong Kong — was transformed dramatically. By the 1890s, Singapore served over fifty regular lines and, unlike ports that served as outlets for hinterlands, the port tapped on its strategic position and became a major regional entrepôt. The rapid development of the port also led to the growth of other ancillary and commercial services, in particular, Chinese businesses in coastal shipping. Discussion on port technology in Singapore before 1940 would also not be complete without reference to the construction of the Singapore Naval Base. It was literally the last bastion of British military and technological supremacy to be unveiled by the British Government in the colony's pre-war history.

By the end of the nineteenth century, Singapore was the pride of the British Empire in Southeast Asia. To reinforce its position as the premier port in the region, Singapore was linked to Britain's imperial telecommunication network. It was a boon to the traders and merchants. Chapter Four examines the technological systems that were gradually introduced to support the rapid urbanization of Singapore and to enhance its living environment. It discusses the impact of imported technologies — telegraphy, modern transportation, electrification, and civil engineering — in transforming the urban landscape. These were transferred with the primary objective of control and to make the environment liveable for the Europeans. The telegraphy satisfied Britain's demand for improved communications between London and the far-flung dominions and colonies at any cost. Steam transport over land and water facilitated the movement of imperial troops and sped the transfer of raw materials to British factories and, in turn, of manufactured commodities to the colonies. Singapore benefited greatly from these technological advances. But, like Hong Kong and Calcutta, rapid urbanization brought along issues of sanitation and health. Particularly from the late nineteenth and early twentieth, sanitation and health became critical issues to be tackled by the colonial administration. Chapter Five discusses the theme of "Technology and the Environment". A major challenge for the municipality in rapidly expanding colonial cities like Singapore, Hong Kong and Calcutta is how to deploy imported sanitation technologies and ideas of environmental aesthetics to prevent disease and promote healthy living within the European community. However, the health of the masses could not be disregarded because Europeans and Asian immigrants alike lived in the shared environment. In any case, the evolution of public health from sanitary measures to environmental health and the wider concerns of social medicine progressed very slowly in nineteenth-century Singapore. Overcrowding, coupled with *laissez faire* policies, lethargy and lack of will by local authorities in the face of apparently insuperable technical difficulties and inadequate financial resources compounded the tasks of the municipal planners and engineers.

While trade was the engine of growth, attempts were also made by Europeans planters and economic botanists to exploit the land for imperial gains. Chapter Six discusses the transfer of colonial scientific ideas in agricultural science and botany to Singapore. Despite the dominance of commerce, a modest start was made in the application of science to the production of tropical export crops during the early decades of the nineteenth century. Colonial botanists, like Henry Ridley, participated in agricultural research of tropical export crops, such as, sugar cane, cinchona, and natural rubber. The Botanic Gardens of Singapore turned the rubber tree into the mainstay of Malayan agriculture. Together with other gardens set up throughout the British Empire, it received a constant stream of information and life plants and by the rotation of personnel, all monitored and planned through Kew. However, it soon dawned upon those who tried to make a fortune out of the land that the island was not endowed with fertile soil. Scientific husbandry and mechanization of agriculture hardly occurred. As it was then and now, Singapore had to depend on imported food. Chapter Seven then documents the role of science and technology in one key feature of the changing pattern of everyday life of the people — food — and seen through the revolutionary incorporation of the Singapore Cold Storage in 1903 and its subsequent growth before 1940. As an agent of change, the company was a powerful engine of innovation and change in food habits and in the development of new food-retailing systems. It imported and retailed "imperial" food catered largely to the European consumers. While a more balanced diet, improved sanitation and modern medical facilities, gave the European population and the well-to-do locals, better health, the masses did not enjoy the benefits of the advances in food technology, "scientific" sanitation and healthcare measures until way into the decades of the twentieth century.

Technology transfer is not just about erecting complexes and monuments in the colonies of imperial powers. It is also about knowledge and learning activity. The perceptions the Europeans had of the learning capacity and culture of their colonial subjects certainly shaped the kind of education they offered. Chapter Eight argues that the upgrading of English and technical education in Singapore during the years 1900 to 1940 was hampered by administrators who were not willing to implement progressive changes. The availability of an English education was limited and, at the same time, vernacular education was actively supported. The perception of the British administrators that Singapore was a trading port and thus had no need for highly trained personnel in scientific and technical professions led to the underdevelopment of a technological culture. Finally, in Chapter Nine, the issue of development of an industrial base and transfer of industrial technology is explored. Despite the extensive application of modern technology to the transformation of the port and the urban environment, by the end of 1940 (and even up to 1965 when Singapore gained its independence), Singapore remained nothing more than a staple port. The development of an industrial and manufacturing base was insignificant and hence, transfer of industrial technology to local businessmen was also capped. Answerable to their shareholders in Britain, British agency houses were not willing to diversify into manufacturing activities that did not really complement their interests in the rubber industry or were in direct conflict with imported products manufactured by their fellow countrymen in Britain. On the other hand, Chinese businessmen, who had successful ventures in the rubber and pineapple industry, were more willing to invest in manufacturing enterprises and to depend on European engineers for technical or managerial support. The rise of these Chinese industrial conglomerates, however, were far and few. It was in the field of Chinese banking that British technology transfer, in terms of knowledge and institutional practices, was most direct and generally successful.

The conclusion revisits the central theme of the book by appraising the relationship between Western science, technology and entrepôt colonialism, as seen in colonial Singapore. The British imparted two outstanding legacies. These were the creation of a world-class port and a modern city. The port was Singapore's economic lifeline during the period of British rule. Its steady growth was due to the benefits of British expertise in port management and shipping technology. While the colonial government failed the acid test of providing sufficient housing for the masses, it gave Singapore a modern infrastructure with good roads, piped sewers, reservoirs, public transport, airports, telecommunication and air services. These technological projects were perceived by British administrators as visible symbols of colonial power and for the population at large, a reminder of the supremacy — and invincibility, at least before the surrender of the "Invincible Fortress" to the Japanese imperial forces — of the British race.

Notes

1. The statue was unveiled by the Governor, Frederick Weld, who in his speech, described Stamford Raffles as "an illustrious administrator and statesman, whose sagacious foresight laid foundations upon which have been built up a great centre of commerce, a focus from which British influence, carrying with it the light of civilization, radiates far around, and which has not only been a blessing to thousands, but has added, directly or indirectly, in no inconsiderable measure to the extent and resources of that vast Empire over which Her Majesty rules". See *Straits Times Weekly Issue*, 6 July 1887.
2. Lee Kuan Yew, *From Third World to First: The Singapore Story: 1965–2000* (Singapore: Times Editions, 2000), p. 67.
3. Donald Moore, *The First 150 Years of Singapore* (Singapore: Donald Moore Press, 1969), p. 15.
4. Quoted in ibid., p. 30. The geographical locale factor for Singapore's successful economic transformation is elaborated in Philippe Regnier, *Singapore: City-State in South-East Asia* (Honolulu: University of Hawaii Press, 1987). Regnier stresses on the complex web of interrelationship

between Singapore and (a) the immediate region and (b) the rest of the world. Situated at a major crossroads of international trade, the city-state's economic success since 1965 was largely due to its ability to continue to function in its historic vocation as an international trading emporium during the period of British rule, and supported by constant upgrading to its modern infrastructure.
5. From the 1700s, the East India Company expanded rapidly on the back of its trading activities and, at its peak, the company employed a third of British workforce and was responsible for 50 per cent of global trade. The company was dissolved in 1874. Today, the 400-year-old brand was acquired and relaunched by an Indian entrepreneur as a high-end, luxury shop in London's upmarket Mayfair district, selling tea, biscuits and chocolates. See also Anthony Webster, *The Twilight of the East India Company: The Evolution of Anglo-Asian Commerce and Politics 1790–1860* (Woodbridge and Rochester: Boydell Press, 2009). Webster examines the intricate commercial and political relationships between the company, the British state and British businesses in finance and industry. He argues that, despite the demise of the company, the operations of the company had led to a flood of capital investment — but dismissing the negative impact on the indigenous Indian industries.
6. See Kwa Chong Guan, *Singapore, a 700-year History: From Early Emporium to World City* (Singapore: National Archives of Singapore, 2009).
7. Peter N. Stearns, *The Industrial Revolution in World History* (Boulder: Westview Press, 1993), p. 83.
8. See, for example, Sheridian M. Tatsuno, *Created in Japan: From Imitators to World-Class Innovators* (New York: Harper & Row, 1990) and Ian Inkster and Fumihiko Satofuka, eds., *Culture and Technology in Modern Japan* (London: I. B. Tauris, 2000).
9. Jonathan Eyal, "History as it should be taught", *Straits Times*, 17 July 2010.
10. See the seminal works of the following authors: C.M. Turnbull, *The Straits Settlements 1826–67: Indian Presidency to Crown Colony* (Singapore: Oxford University Press, 1972); Edwin Lee, *The British as Rulers: Governing Multicultural Singapore, 1867–1914* (Singapore: Singapore University Press, 1991); Wong Lin Ken, "Singapore: Its Growth as an Entrepôt Port, 1819–1914", *Journal of Southeast Asian Studies* 9 (1978): 50–84.
11. Two recent works are Sarah Stockwell, ed., *The British Empire: Themes and Perspectives* (Oxford: Blackwell Publishing, 2008) and Radhika Mohanram, *Imperial White: Race, Diaspora, and the British Empire* (Minneapolis and London: University of Minnesota Press, 2007).
12. This book adopts Wikipedia's definition of the term "technology transfer":

"Technology transfer is the process of sharing of skills, knowledge, technologies, methods of manufacturing, samples of manufacturing and facilities among governments and other institutions to ensure that scientific and technological developments are accessible to a wider range of users who can then further develop and exploit the technology into new products, processes, applications, materials or services". It is not merely the physical introduction into the receiving country of various kinds of supposedly more efficient Western hardware. It involves transmission of the complex of knowledge, materials, and methods pertinent to the operation of the technics or technical systems in question.

1
Technology and the British Empire

The Industrial Revolution took place in Great Britain at about the 1760s, and by the 1850s Britain was crowned "the workshop of the world". To celebrate, Queen Victoria opened "The Great Exhibition" at Crystal Palace on 1 May 1851.[1] It was opened to all nations — including the "uncivilized" nations — and the main purpose was the display of Britain's industrial achievements. *The Times* reported on 2 May 1851 that it was "a sight the like of which has never happened before, and which … can never be repeated…".[2] The Queen herself recorded in her journal:

> What used to be done by hand and used to take months doing is now accomplished in a few instants by the most beautiful machinery. We saw first the cotton machines from Oldham… We saw hydraulic machines, pumps, filtering machines of all kinds, machines for purifying sugar — in fact, every conceivable invention. We likewise saw medals made by machinery which not more than fifteen years ago were made by hand, fifty million instead of million can be supplied in a week now.[3]

The Industrial Revolution was a period blessed with technological triumphs that did not draw upon theoretical knowledge.[4] Many of the machines invented had little to do with the science of the day. The rise of the textile industry, which accounted much for the economic growth of eighteenth century Britain, was not the result of the application of scientific theory. The inventions of John Kay, James Hargreaves, Richard Arkwright and Samuel Crompton were essentially owed more to past craft practices than they did to science. They were the outcome of evolutionary changes within technology.[5] The arrival of modern science only began to make its mark from the latter half of the nineteenth century. The Great Exhibition of 1851, and similar ones that followed, glorified industrial progress and the men who made it possible. More significantly, they were also used to measure the relative industrial growth of nations. The accelerating changes brought about by the machines, and the steam engines that powered them were truly revolutionary in the ways they affected the lives and fortunes of the people of Great Britain. Technology became a factor in international affairs and conflicts. As the "workshop of the world", Great Britain led other European nations — Germany, Austria, Hungary, Sweden, Italy, Russia, the United States (and Japan) — in creating an empire during an era of "new imperialism", popularly denoted as the decades from 1870 to 1914. The victims were the densely populated and ecologically inhospitable countries in Africa and Asia. Niall Ferguson terms it as "Anglobalization with gunboats" — which would eventually made the British Empire the biggest empire ever created by the ingenuity of man.[6] In his seminal study of environmental history, Michael Williams points out that the rise and expansion of industrial Europe, particularly the period from 1850 to 1914, was characterized by the "explosive increase of European population and its movement overseas, and the rise of modern capitalist economy and its evolution into industrialism".[7] No other European countries exemplify these phenomena more intensely than Britain — the pre-eminent industrial and imperial economy of the "Age of Empire". By 1914, the British Empire consisted of nearly a quarter of the earth's population and land mass.

A voluminous amount of literature has been written on the motives behind the great expansion of empires in the last quarter of the nineteenth century. In the late 1930s, J.A. Hobson put forth his "metropolitan" interpretation where the motivations and policies of the British are examined with particular reference to economic forces and political decisions emanating from Britain. Hence, the partition of Africa was deliberately planned by London financiers, investors and politicians to exploit the economic resources and manpower. The argument which raged with much vigour in the 1960s and 1970s centres on the classic Robinson-Gallagher's "peripheral" thesis. Robinson and Gallagher argue that the British really preferred to have "informal control", that is, economic influence without the expense of political control. While they did not place much emphasis on the economic dimension, Robinson and Gallagher put forward the hypothesis that Britain had always been concerned to protect her overseas trade. Only the methods, such as the use of collaboration, changed with changing circumstances. The debate continued with the work of C.D.M. Platt, Bernard Semmel, and David Fieldhouse, all giving greater attention to trade as a trigger factor of British imperialism. In the late 1980s and within a growing interest on a reappraisal of the dominance of different groups in the British society, Peter Cain and A.G. Hopkins stimulated the imperialism debate to a new height by offering a new interpretation of the forces behind British imperial expansion. They singled out the role and motivations of the "metropole" (particularly, London) service sector dominated by "gentlemanly capitalists" such as bankers, insurers, and non-aristocratic businessmen, as the ones who drive the expansion of the British Empire through investment, commerce and migration in the "periphery".[8] Whatever the motives of the imperialists, the penetration of Africa and Asia was done with swiftness and thorough success, in contrast to their efforts at the beginning of the century. Some historians, particularly Daniel Headrick, attributes this robustness of expansion to the advancements made in science and technology in the Western, industrialized countries.[9] Departing from the more established economic and political causes of imperialism, Headrick's major argument in his early contribution to the scholarship on European imperialism is that advances made in science and technology served as a powerful motivational cause for territorial expansion by the European imperialists, especially the British. He argues that the new tools "of the nineteenth century ... shattered traditional trade, technology, and political relationships, and in their place they laid the foundations for a new global civilization based on Western technology".[10]

For Great Britain, it all started as early as the mid-eighteenth century. As part of Britain's preparation for the Seven Years War (1756–63), William Pitt secured a commitment from the British Parliament to recruit 55,000 seamen and increased the fleet to 105. In the process, the Royal Dockyards became the largest industrial enterprise in the world. Not only technology but also science was used to rule the waves. By the time Captain James Cook's *Endeavour* sailed for the South Pacific in 1768, the chronometer and sauerkraut as an anti-scorbutic "epitomized the new alliance between science and strategy".[11] The Europeans' vulnerability to tropical diseases, such as malaria, was also greatly reduced by the use of quinine prophylaxis. Although Europe eventually conquered Africa because of superior military and medical technology, "Europeans have never adapted biologically to the diseases of tropical Africa and must constantly take precautions such as anti-malarial drugs and mosquito nets to preserve their health".[12] Finally, improvements in firearms broke down hostile resistance from indigenous populations.[13] The Gatling guns, Maxims and light artillery were the hallmarks of the "firepower revolution" and effectively contributed to the imperial ambitions of Western nations. Operated by a crew of four gunners, the 0.45 inch Maxim gun could fire 500 rounds a minute, fifty times faster than the fastest rifle available in the late nineteenth century. In the 1880s, Frederick Lugard used the Maxim to its fullest ferocity as he carved out Britain's West African empire. As the High Commissioner for northern Nigeria in 1900, Lugard refined his method of indirect rule, recruiting local chiefs as "collaborators of colonialism". The imperialist justified the Britain's rule of foreign lands by stating:

> As Roman imperialism laid the foundations of modern civilisation, and led the wild barbarians of these islands along the path of progress, so in Africa to-day we are repaying the debt and bringing to the dark places of the earth, the abode of barbarism and cruelty, the torch of culture and progress, while ministering to the material needs of our civilisation... British rule has promoted the happiness and welfare of the primitive races... We hold these countries because it is the genius of our race to colonise, to trade and to govern.[14]

While Lugard's system of rule proved to be largely ineffective in Africa because the lands were fragmented with tribal kingdoms owing allegiance to their chieftains, indirect rule was successfully introduced into the Malay States of the Malay Peninsula where there was distinct local authorities — in the hands of the sultan and his hierarchy of Malay village headmen. The Maxim continued its destructions in Africa when Cecil Rhodes won the Battle of Shangani River against the Matabele tribe in 1893. More than 1,500 Matalebe warriors died and only four of the 700 invaders perished. To the tribal folks the victory was due to the witchcraft symbolized by the evil spirit found in the Maxim. The conquered territory is modern Rhodesia (present-day Zimbabwe). In the words of Ian Morris: "Technology transformed colonization".[15]

THE POWER OF WESTERN TECHNOLOGY

As early as the 1830s, British colonial administrators, missionaries and businessmen came to view Western technology — largely embodied in the railroads, steamships and machines in general — as powerful civilizing artefacts. According to Robert Kennedy, the "impact of the western man", with his knowledge of modern science and technology, on the non-European world was far-reaching and effectively altered the dynamics of world power in the nineteenth century.[16] The impact was manifested not only in "a variety of economic relationships — ranging from the 'informal influence' of coastal traders, shippers, and consuls to the more direct controls of planters, railway builders, and mining companies — but also in the penetrations of explorers, adventurers, and missionaries, in the introduction of western diseases, and in the proselytization of western faiths".[17] Among the specific kinds of cultural contacts that led to the diffusion of technology, imperialism and conquest are of great importance. Under these circumstances, the conquered and backward races were totally dependent on their imperial masters for modern machines, engineering and business skills, and capital investment in order to transform their primitive societies. After 1875, the era of intensified imperialism saw Western technological innovation and diffusion to colonies escalated rapidly. Together with the diaspora of British engineers, colonies in the periphery of the British Empire were gifted with railways, telegraphy, electricity, ports, canals, roads, waterworks, and urban services. These technological tools of imperialism were powerful reminders to indigenous peoples in the colonies of the Europeans' ability to control time and space, to innovate and to create more technological marvels. Though linked to aspirations to improve the livelihood of the peoples, the transfer and diffusion of modern scientific knowledge and technology in Africa and Asia was, more often than not, couched in terms that heightened the Europeans' dominance. Headrick identifies a two-stage technology transfer process.[18] The first stage is the "geographic relocation" of equipment and methods, along with the experts to operate them and the second stage is the "cultural diffusion" of technology during which imported scientific and technological knowledge is assimilated by the recipient culture. Using examples of transfer of some key technologies, such as railway, shipping and telecommunications to colonies, Headrick concludes that European imperialists were successful in relocating Western technology into their colonies but, either intentionally or not, failed to produce a cultural diffusion of the knowledge and skill. The cause of this failure lies in the "unequal relationship between the tropical colonies and their European metropoles [and] in order to obtain the full benefits of Western technology through its cultural diffusion,

Africans and Asians had first to free themselves from colonial rule and then — a more arduous — learn to understand, and not just desire, the alien machinery".[19] Nevertheless, importation of Western technology did provide the opportunities for local collaborators to adapt the imported technologies, such as the steamship, for their own advancement.[20] In the long run, the miniaturization and mobility of many modern weapons and weapon systems made it impossible to prevent their diffusion to the colonized peoples. Moreover, mass communication technological systems facilitated the crystallization and growth of senses of identity and oppression among colonized peoples. Hence, ironically, technology eventually played an important part in the demise of political colonialism.

GREAT BRITAIN, THE LEADING IMPERIALIST

In the century between 1820 and 1920, 55 million Europeans (or one-fifth of the population in Europe in 1820) moved to the lands colonized by their respective imperial governments. The British formed the bulk of this "Anglobalization" process of creating colonies, led by agents of change and supported by financial backing, firepower and technology transfer. It was not just "Civilization, Commerce and Christianity". "Conquest" was now added to complete the jigsaw puzzle of the British Empire. Within a short period of about two decades after 1880, ten thousand African tribal kingdoms were subjugated and transformed into just forty states, of which thirty-six were under direct European control. The scramble was an expression of European rivalries and international balance of power. By the start of the First World War in 1914, apart from Abyssinia and Liberia, the entire African continent was under some form of European imperialism, and about a third paid homage to Great Britain. This was what came to be known as "the Scramble for Africa" — although Ferguson candidly affirms that "the Scramble *of* Africa might be nearer the mark". And in the Far East, another scramble for land, extraterritoriality, power and control was taking place in the once mighty Middle Kingdom of China, north of Borneo, Malaya and Singapore, a greater part of New Guinea and a chain of islands in the Pacific. In 1860 the territorial extent of the British Empire had been some 9.5 million square miles; by 1909 it had risen to 12.7 million, covering 25 per cent of the world's land surface. Within this titanic land mass, some 444 million people potentially stood to benefit from the "civilizing mission" of British imperial rule.

No other power had invested more in their colonial empires than Great Britain during the period 1870 to 1914. An important aim of imperial rule was the promotion of colonial prosperity. "Civilization", often defined by the Victorians in terms of "improvement", "progress", and "regeneration", had an economic dimension. While economic historians continue to probe the role of the British government in the economic development of Britain's tropical and sub-tropical dependencies, the general consensus is that the metropolis economic involvement in these dependencies was significant. Richard Kesner concludes: "The British government did not so much control the economies of her Crown Colonies as participate in their economic development. This participation took on various forms: some of them direct, such as public works construction, and some indirect, such as the supervision of colonial banking and currency services. Throughout, the overriding imperial objective was the creation of economically strong and self-reliant colonies tied to Britain by choice and sympathy rather than by compulsion."[21] Rising in the 1850s and 1860s, Britain's net foreign investment averaged about a third of the nation's annual accumulations from 1870 to 1914.[22] India was regarded as the centre of the British Empire and, indeed, the envy of all other colonial powers. Of the 1,789 million pounds in long-term publicly issued British capital invested in the Empire up to 1914, India and Ceylon received 379 million pounds, or 21.3 per cent.[23] While the issue of costs and benefits of this immense capital outlay still rages on among scholars, what is certain is that the period saw a massive flow of technology and skill from Britain to her colonies, especially India. In return, agricultural produce, such as indigo, gutta percha, palm oil, tin and rubber from the newly acquired tropical colonies were shipped in bulk to benefit British industries and society. Increasing amounts of sugar, tea, coffee and other tropical goods satisfied the daily needs of the British consumers' markets.

Economic expansion in the tropical periphery was based on the expansion of both plantation and peasant production, backed up by imported capital and infrastructure, such as the construction of railways. Unlike the old plantation colonies located mainly in the Caribbean where the small white settler landlords and plantation owners dominated the economy serviced by ex-slaves of African descents, Malaya and Singapore were the newer plantation colonies where the labour force was largely indentured from India and China and considerably more British capital and technology was invested in them. Some of these tropical colonies, like

Singapore, were advantageously-placed bases along the sea lanes of the world, with good harbours and defensible positions. In general, Britain and the other European powers set out to create a neo-European social, ecological, and economic setting in their tropical possessions, based on British law and justice, settled agriculture, and primary product extraction activities. However, the introduction and impact of industrial capitalism into the tropical periphery was less intense than in the neo-Europes (the colonized temperate lands of the Americas, Australia and New Zealand) because the environment in the tropics was less responsive to European technology.[24] British investors readily transferred capital and technology to the "neo-Europes" and supplemented by large influx of British emigrants, particularly those with engineering and technical knowledge and skills. On the other hand, for the tropical periphery, climatic and topographical difficulties, inadequate knowledge, institutional resistance to change and peculiar indigenous social, political, and economic systems were significant inhibitive factors to the spread of industrial capitalism. There was one exception — Japan — where internal conditions during the Meiji era were favourable to the adoption and diffusion of European technology and knowledge, resulting in the establishment of a self-sustaining growth process.[25] It is argued here that, while the inflow of British industrial technology and expertise was far and few, the periphery port colony of Singapore had intimate business ties to Britain and relied heavily on British exports of capital, manpower, and trading enterprise to expand its economy. This imperial connection became more focussed and, indeed, Britain became more dependent on her Empire as other European nations industrialized and compete with her for economic supremacy.

As Britain expanded her influence into lands way beyond the shores of Europe during the course of the nineteenth century, her statesmen had to constantly define, review and redefine policies towards her tropical colonies. There was what might be called the "unofficial orthodoxy" of the Manchester school of *laissez-faire* dogmatism, anchored strongly on the concept of free trade and informal rule, and backed by naval power that reinforced the idea of "the freedom of the sea". In his highly charged essay "Chartism", Thomas Carlyle extolled his countrymen that, for the sake of progress of England, it was their moral obligation to exploit untapped resources of lands beyond. There was the assumption held by the Europeans that it was both natural and mutually beneficial for advanced European societies to provide the natives with machines and manufactured goods in return for primary products exported from colonized lands. For Britain, this was seen as a logical extension of its inauguration of, what Richard Rosecrance described as, a "non-territorial model of international politics", an entirely new commercial system based upon free trade beginning in the 1840s.[26] The argument was that if uninterrupted international trade could be extended to her markets which supplied her raw materials and food, Britain then need not resort to territorial acquisitions to gain economic benefits. Under the law of comparative advantage, an open-trading system allowed Britain to concentrate on pursuing an industrial and export economic policy while her colonies and dominions specialized in the production and supply of cheap raw material. At the same time, imperial bondage ensured that Britain's colonies purchased British-made goods through their agents in London.

Within the tropical colonies, the hands of the British statesmen in London were frequently pushed into more direct intervention of domestic affairs by the aggressive activities of the man-on-the-spot, the British administrator, trader, the financier, the consul, the missionary, and the naval officer, each with his own vested interest. They deployed, sometimes forcefully, the indigenous peoples to carry out unregulated extraction of natural resources — using their superior knowledge of science and technology, entrepreneurial and technical skills. These British agents of change were, after all, upholding Britain's "imperialism of free trade". From the imperial government's point of view, the role of colonial administration was to establish and maintain a British presence in the colonies, to provide certain basic services, such as law enforcement and mail delivery, and to raise the necessary revenues to pay for its own operation. However, if the colony's resources were not sufficient to bear the burden of a particular project, it was generally shelved until there was sufficient funds. Appointed in 1867 as the first colonial Governor of Singapore, Henry Ord was instructed to rule efficiently and keep within the allocated budget. Crown Colonies relied upon indirect taxation for most of their revenues. For the duty-free ports of Singapore and Hong Kong, taxes were levied on the sales and consumption of a limited number of items. This framework allowed for considerable discretion on the part of the colonial administrators. They had to deal with "the dilemma of either preserving indigenous institutions in order to promote stability or Anglicising those societies in order to modernise their activities" and their administrative philosophy "also fluctuated over time as political ideas and cultural fashions changed within Britain itself".[27] Even when communication between London and the colonies was improved through steamship and telegraphs, the fundamental bureaucratic problem remained, that is, the need for continuous adjustment for translating imperial agendas formulated in London in the colonies with local beliefs and customs. All said, the local officials played a decisive role in determining the strategies and directions of imperial control and exercised considerable latitude of authority — sometimes to the embarrassment of London, as was the case of Sir Andrew Clark, governor of the Straits Settlements, who exceeded the instructions given to him in London by promulgating the Pangkor Engagement in January 1874 which opened the door to the eventual British domination in the Malay Peninsula.

The end of the nineteenth century and the beginning of the twentieth is often designated as the "High Noon" of the British Empire. Although the Empire was at its greatest geographical extent only after 1918, this period was a time when popular enthusiasm for imperialism was at its height, and when ideas about empire most thoroughly saturated domestic British culture. At the "core", colonial development policies were inextricably linked to Joseph Chamberlain. As the Colonial Secretary from 1895 to 1903, Chamberlain was seen by historians as ushering in a period of "constructive imperialism" and the concept of "indirect rule" emerging after 1900 as a much publicized technique of Imperial management of the tropical colonies.[28] After taking office in June 1895, he told the House of Commons that the:

> underdeveloped estates could never be developed without Imperial assistance ... cases have already come to my knowledge of colonies which have been British Colonies perhaps for more than a hundred years in which up to the present time British rule has done absolutely nothing ... I shall be prepared to consider very carefully myself, and then, if I am satisfied, confidently submit to the House, any case which may occur in which by the judicious investment of British money those estates which belong to the British Crown may be developed for the benefit of their population and for the benefit of the greater population which is outside.[29]

Chamberlain's "judicious investment" took the form of steamer and cable subsidies, railway, roads and harbour constructions, irrigation and drainage, sanitation works, health programmes, and the introduction of British practices in banking, insurance and fiscal reforms. Through his persistent effort, the British government passed the Colonial Stocks Act of 1900 by which the colonies were given the privileges of being empowered to issue loan stock with trustee status, which was tantamount to an imperial guarantee. Under this arrangement, private investors demonstrated confidence in newly issued Crown Colony stocks and in the years prior to 1914, colonial efforts at fund raising for development projects increased. For the Straits Settlements, the loans issued by the Crown Agents increased from a mere 100 for the period 1870 to 1879 to 5,000 for the period 1900 to 1909.[30] While some historians would stress the limitations of Chamberlain's achievements, there is no doubt that the Colonial Office adopted a proactive stand towards development in the tropical colonies in the new century prior to the outbreak of the First World War in 1914. The world economy was on the upswing at the turn of the century and tropical colonies like the Straits Settlements, Malaya, Ceylon, Nigeria, Trinidad and Tobago, and the Gold Coast enjoying surging economic growth. Malaya, including the Straits Settlements, was the star performer, leading the tropical colonies with a revenue per head of 2.32 million pound in 1911, as compared to a total of 3.26 million pound per head for the next four colonies (Ceylon, Nigeria, Jamaica and Gold Coast) with the highest revenues in 1911.[31] Nevertheless, the Colonial Office's stood firm on its development philosophy, that once the colonial administration had provided a framework of ordered government and basic infrastructure, private entrepreneurs and private capitals could be depended upon to initiate further development projects.

"CIVILIZING MISSION"

The British also believed in white supremacy and promoted humanitarian ideas. The literature on "civilizing mission" — and the assumptions about European technological and scientific superiority in which it was rooted — as a serious ideology is not extensive even though the ideas associated with it were pervasive in late nineteenth-century European writings on imperialist expansion in Africa and Asia. Indeed, in the era of decolonization, nationalists, politicians, and even historians tend to dismiss "civilizing mission" as a moral explanation by the Europeans of their conquest and exploitative rule. Michael Adas explains:

> Undoubtedly, claims that colonial conquests had been undertaken in order to uplift African and Asian peoples could be little more than cynical camouflage for brutal exploitation … But many of those who justified imperial expansion or colonial policies in the name of higher purposes linked to the civilizing mission were firmly convinced that they were acting in the long-term interests of the peoples brought until European rule.[32]

Adas's extensive research on the role of scientific thought and technological innovation examines how these elements created an ideology of Western superiority with regard to non-Western peoples. He argues that this ideology of Western superiority is an integral part of the imperial justification for both the paternalistic civilizing mission and the rapid spread of European hegemony. Stephen Howe elaborates:

> Empire depended, in the sphere of ideology, on ideas about difference, and usually, on a belief in superiority. In earlier eras the form of such ideas had been predominantly religious. By about 1900, though these had not disappeared, the most powerful notions were cultural, civilizational, and — perhaps more than at any time before or since — racial.[33]

The notion of "colonial modernity", largely propagated by cultural historians of empire, has been widely invoked. It expresses on the idea of colonialism as a "modernizing, state-building, centralizing, developmentalist, and secularizing force".[34]

Social and cultural historians have consistently highlighted the interlocking relationship between mission, power and colonialism. Anchoring this relationship is the missionaries' frequent invocation of European superiority, including the resolute belief that Christianity was superior to all other religion and the path to salvation of pagan peoples. Alison Twells, in her work on the importance of missionary philanthropy for middle-class formation in Britain and the roles of men and women missionaries as "handmaidens" of British imperialism in Ireland, West Indies, West Africa and the islands of the South Pacific, concludes:

> By 1850, the national civilising mission drew upon Biblical, commercial and secular civilising technologies, involved the wider English populace and was sustained from within popular middle class as well as evangelical culture. The cultural ambitions of the early-nineteenth-century middle class thus paved way for the imperial ambitions of the nation in mid-nineteenth-century England, as the civilising mission was consolidated

as a national project. Missionaries were now at the centre of the master narrative of imperial history and at the heart of the cultural life of the nation.[35]

In the far corners of the British Empire, British missionaries, missionary doctors, and religiously motivated volunteers, described by Sheldon Watts as "a powerful interest group which colonial administrators ignored at their peril"[36] undertook health work, particularly the fight against leprosy — popularly regarded as a God's punishment for dark, hidden thoughts, words and deeds. Like their military and administrative counterparts, British missionaries' attitudes towards non-Western peoples and their own sense of righteousness over the "heathens" were shaped by their superiority in scientific knowledge and technical skills and that Christian civilization had provided for the emergence of the scientific and technological revolution. David Livingstone, William Carey, Abbe Boilat, John Cumming and others underscored the importance of the scientific and technological connection to efforts to destroy the bondage of superstition and spread Christianity.[37]

The more fundamental and long-term implication of the civilizing mission ideology centres on, in Adas's words, the "attitudes toward non-Western peoples and cultures which were held by literate members of the upper and middle classes of European societies, and the ways in which these attitudes shaped ideologies of Western dominance and informed colonial policy-making".[38] In the late eighteenth century, the motivations behind British superiority were anchored on morality, law, religion and political institutions. With their long history and proud traditions, the British viewed the regimes and hierarchies of the lands in Africa and Asia as backward, inefficient, despotic and corrupt and had to be replaced by the more civil and advanced model of Western society and political institutions. The belief was that these backward societies were capable of eradicating barbaric practices and political and social tyranny. In the process, they could live a civilized and orderly existence and seek enlightenment in Christianity. However, as the British Empire expanded in the nineteenth century and as the gulf of progress between Britain the metropolis and the colonies in the periphery widened, it was obvious that Africans and the Asians did not have the capacity to reform and achieve the great advance anticipated. The Indian Mutiny of 1857 was a substantial confirmation of the need for Britain to rule and to control. The British Empire took a racist stance — the British were inherently superior, and that other races were inferior and those races therefore required different treatment. In the nineteenth century, race became a social scientific tool to explain and stereotype peoples of diverse characteristics, but also levels of developments. The British saw themselves as teachers of the backward and superstition-bound peoples in Africa and Asia. Alfred Milner, the High Commissioner for Cape Colony and Cecil Rhodes, the owner of diamond and gold mines in South Africa and members of what George Young has aptly described as the "articulate classes", are par excellent examples of white imperialists notorious for their blatant expressions of their racial supremacy.[39] Sharing the same views as the Colonial Secretary Joseph Chamberlain, Milner advocated that colonies should be closely bonded to the Empire and that this imperial consolidation was strictly racial and class based. He commented in *The Times*: "My patriotism knows no geographical but only racial limits. I am an Imperialist and not a little Englander, because I am a British race patriot".[40] Rhodes reinforced the white supremacy ideology by stating that the English were the "chosen instrument to carry out the divine idea over the whole planet … we are the finest race in the world and the more of the world we inhabit the better it is for the human race".[41] The Victorians regarded themselves as leaders of civilization and as pioneers of industry and progress. It was their destiny to develop the resources of the globe through imperial expansion and their duty to propagate the civilizing mission ideology. Similarly, American politicians, social commentators, and missionaries viewed the country's exceptional technological and material achievements as proof of its "Manifest Destiny" to take the lead in the quest of "civilized" Westerners to uplift the peoples of Asia.[42] For the Americans, they attempted to bestow their civilizing ideology onto the people of the Philippines; for the British, India was their exemplar model for success in their civilizing mission effort and for other colonies in the empire to envy and emulate.

THE CASE OF INDIA

The Seven Years War had irrevocably made India the prized possession of the British — and not the French — with both a huge market for British trade and an inexhaustible reservoir of military manpower. India was more than the "jewel in the crown" — it was "a whole diamond mine".[43] During the two hundred years when they governed India, the British introduced virtually every aspect of their material culture to the subcontinent. Western science and technology were transferred with a political purpose of consolidating British rule and control of the continent.[44] Although the indigenous people were not placed in command of these new technological innovations, they were involved in peripheral occupations that at least allowed them to appreciate modern machinery and Western ideas of novelty, change and progress. Moreover, as argued by Zaheer Baber, through the expansion of English education and the establishment of engineering and technical schools, Indian scientists were trained and were playing a more active role in the development of scientific institutions in India.[45] Zaheer reiterates that while British scientific and technological supremacy did contribute to colonial expansion and control in India, "colonial rule itself led to the creation of new forms of knowledge and institutions that were replicated in Britain and elsewhere".[46] Writing in 1894, British sociologist, Benjamin Kidd, provided the most resounding late Victorian affirmation of the role of technology in Britain's civilising mission in India:

> There has been for long in progress in India a steady development of the resources of the country which cannot be paralleled in any other topical region of the world. Public works on the most extensive scale and of the most permanent character have been undertaken and completed; roads and bridges have been built; mining and agriculture have been developed; irrigation works, which have added considerably to the fertility and resources of large tracts of the country, haven been constructed and even sanitary reform is beginning to make considerable progress. European enterprises too, attracted by security and integrity in the government, has been active.[47]

But scholars also argued that today India is often cited as an example of economic development went wrong and as an example to illustrate the broad, negative impact of imperialism on the economic development of the developing countries. Many explanations were offered for the failure of India to experience its own "Industrial Revolution". Some blamed it on the exploitation of its resources by the Europeans. David Landes attributes India's failure to the Indians themselves. This is because "no one seems to have had a passionate interest in simplifying and easing tasks [and] both worker and employer saw hard labour as the worker's lot — and as appropriate".[48] Landes cites the railway construction where, even with the introduction of modern, labour-saving machinery, Indian workers would rather "move earth and rock by hand".[49] And, with the wheelbarrow, it was reported that Indian labourers would place it on their heads rather than wheel them! He explains:

> [A]lthough the colonialists often left behind an infrastructure of roads, ports, railroads, and buildings, maintenance was another matter. The ability of the ex-colonies to neglect and run down their material legacy was stunning ... Much of what these subject populations learned in the schools and universities of the colonial master was political and social discourse rather than applied science and technical know-how — the makings of revolution rather than production.[50]

Gregory Clark argues that the critical reason for India's backwardness was in the inabilities of the Indians to use the new technologies effectively — as seen in the factory production of cotton textiles and railways.[51] Inkster, in his study of the impact of Western science and technology on the economic development of imperial colonies, also categorised India as a "loser". Inkster remarks that "[g]rowth in India was enclavist".[52]

By the mid-twentieth century, the British Empire was effectively gone. At its apogee, it was one of the largest territorial empires the world has ever witnessed, and it profoundly shaped the lives of people both in Britain and overseas. The debate as to whether former colonies of Britain — and the other European empires — came out as "winners" or "losers" is still a rallying issue for discussion. Undoubtedly, some imperial nations were better rulers than others and their colonies performed better after gaining full sovereignty and independence. For the Spaniards, the conquest of the Americas was accomplished with much cruelty and treachery, and all in the name of seeking and controlling the treasures of the lands. For the Portuguese in Asia, fortresses and defensible strongholds such as Goa had to be built in order to control trade, the local merchants and the population at large. The Dutch and the English sought trade, although "Christianity, commerce and civilization" often led to embroilment in local affairs and subsequent conquest of the lands. Historians generally agreed that the British were the most benevolent of the European imperialists because of their willingness and ability to invest in social overhead and their reliance on local elites to administer their "informal" empire. "What the British Empire proved", writes Niall Ferguson, "is that empire is a form of international government that can work — and not just for the benefit of the ruling power — [and] the notion that British imperialism tended to impoverish colonized countries seems inherently problematic".[53] In *Civilization*, the historian supports his stand that Western dominance has been a good thing for the world.[54] Ferguson argues that the West ascendancy is based on six attributes which he labels as "killer apps": competition, science, democracy, consumerism, the work ethic and medicine which he considers the most remarkable. For Great Britain, the spread of British rule between the 1840s and 1930s gave to the world the institutions of parliamentary democracy, a global network of modern communications, the rule of law and, especially during the nineteenth century, free trade, free capital movements and, with the abolition of slavery, free labour. Finally, though Britain fought many small wars, the imperial army had maintained a global peace for a long period of time. Britain encouraged its imperial investors, traders and merchants to put their money in wide-flung colonies where British rule had reduced investment risks. Within the peace and order of the *Pax Britannica*, the British Empire became a gigantic development agency, distributing technical knowledge around the world, and erecting infrastructures of industrial progress — roads, railways, ports, posts and telegraphs — methodically carried out by British agents of change.

Notes

1. The "Crystal Palace", erected in London's Hyde Park in 1851 to house the exhibition, was the creation of Joseph Paxton, a landscape gardener. It demonstrated what could be done by a judicious combination of over a million feet of glass and iron. Although intended to be a temporary structure, the Crystal Palace was so popular that it was later re-erected in South London where it stood until 1936 when it was accidentally destroyed by fire. Fast-forward to the twenty-first century — in Shanghai, China where 189 countries and 57 international organizations gathered for The World Expo in May 2010. For Shanghai, the milestone event marked its rise as a world-class city and for China, its rise as a global power.
2. *The Times*, 2 May 1851. The prevailing attitude in England at the turn of mid-nineteenth century was ripe for the somewhat arrogant display of her accomplishments and those of her colonial subjects. Many living in England felt secure, economically and politically, and Queen Victoria was eager to reinforce the feeling of contentment with her reign. It was during this time that the word "Victorian" began to be popularly used to denote a

new self-consciousness and confidence, both in relation to the nation and to the period through which it was passing.
3. Eric de Mare, "The Year of the Great Exhibition", quoted in *From Waterloo to the Great Exhibition: Britain 1815–1852*, edited by Colin McNab and Robert Mackenzie (Edinburgh: Oliver & Boyd, 1982), p. 173.
4. In recent years, the British Industrial Revolution has come under serious scrutiny. Some scholars have argued that there was little economic growth in Britain in the second half of the eighteenth century and, hence, the Industrial Revolution must not be identified with continuous economic growth. Others have suggested that it is more appropriate to think in terms of accelerating and unprecedented technological change, spurred on by novel ideas and imaginations that made it possible to produce more and quality goods and do so more efficiently. A recent addition to the scholarship is by Gregory Clark. In his book *Farewell to Alms*, Clark asked why did the Industrial Revolution happen in the late eighteenth century England and not at some other time or place and what has prevented some nations from fully capitalizing on the benefits of industrialization. See Gregory Clark, *A Farewell to Alms: A Brief Economic History of the* World (Princeton: Princeton University Press, 2007). See also Joel Mokyr, *The Lever of Riches: Technological Creativity and Economic Progress* (New York and Oxford: Oxford University Press, 1990) Chapter 10; Philip Woodine, *The Industrial Revolution* (University of Huddersfield: Huddersfield Pamphlets in History and Politics, 1993); David Landes, *The Wealth and Poverty of Nations* (London: Little, Brown and Company, 1998), Chapter 15, and David Cannadine, *Making History Now and Then: Discoveries, Controversies and Explorations* (New York: Palgrave Macmillan, 2008), Chapter 4. For a recent scholarship on why the Industrial Revolution happened in the West (in Britain) and not in the East, see Ian Morris, *Why the West Rules — For Now: The Patterns of History and what they reveal about the Future* (London: Profile Books, 2010), Chapters 9 and 10.
5. The musket and the cannon are examples of two artefacts that epitomize the evolutionary improvement in technology through the centuries and gunpowder was made progressively more efficient. The first recorded mention of the cannon was made in 1318 and in due course smaller cannons appeared and eventually the flintlock muskets were developed. See George Basalla, *The Evolution of Technology* (Cambridge: Cambridge University Press, 1988); reprinted 1989, 1990. Basalla challenges the popular notion that technological advances by the efforts of a few individuals who produce a series of revolutionary inventions owing little or nothing to the technological past.

6. Niall Ferguson, *Empire: How Britain Made the Modern World* (London: Penguin Books, 2004), p. xxiv.
7. Michael Williams, "The Relations of Environmental History and Historical Geography", *Journal of Historical Geography* XX (1994): 12.
8. See C.D.M. Platt, "Economic Factors in British Policy during the 'New Imperialism'", *Past and Present* 39 (1968): 120–38, B. Semmel, *The Rise of Free Trade Imperialism* (Cambridge: Cambridge University Press, 1970) and David Fieldhouse, *Economics and Empire, 1830–1914* (London: Weidenfeld and Nicholson, 1973). Cain and Hopkins's comprehensive new interpretation is covered in their recently published, *British Imperialism*, Vol. 1, *Innovation and Expansion, 1688–1914*, and Vol. 2, *Crisis and Deconstruction, 1914–1990* (Longman, 1993). For a succinct, easy-to-follow introduction to the key issues and historiography of British imperialism, see Robert Johnson, *British Imperialism* (New York: Palgrave Macmillan, 2003), Chapter 4 and Andrew Porter, *European Imperialism, 1860–1914* (London: Macmillan, 1994).
9. See Daniel R. Headrick, *The Tools of Empire: Technology and European Imperialism in the 19th Century* (New York: Oxford University Press, 1981); and *The Tentacles of Progress: Technology Transfer in the Age of Imperialism 1850–1940* (New York: Oxford University Press, 1988). In his latest contribution, Headrick expands his "imperialism and technology" theme to cover the rise and expansion of Western imperialism in the past 600 years and sheds light on the environmental and social factors that influenced the successes and failures of Western powers in using technology in empire-creation. See Daniel R. Headrick, *Power Over Peoples: Technology, Environments, and Western Imperialism, 1400 to Present* (Princeton: Princeton University Press, 2010).
10. Headrick, *Tools of Empire*, p. 177.
11. Ferguson, *Empire*, p. 33. On board the *Endeavour* was a group of naturalists, including the botanist Joseph Banks.
12. David Clark, *Germs, Genes and Civilization: How Epidemics Shaped Who We Are Today* (New Jersey: Pearson Education, 2010), p. 122.
13. Headrick, *Tentacles of Progress*, p. 6. In the case of Britain's "conquest" of India, it can be argued that superior armament is just one of the vital success factors. The Indian armies too were well equipped. More significant was the gradual collapse of the Mughal government, the good discipline of the British forces and bold strategy of commanders like Arthur Wellesley, the future Duke of Wellington. Wellesley's contribution also illustrates the distinctive Western approach to the organisational aspect of technology. Indian armies might have had good weaponry, but because their guns were

made in great variety of shapes and sizes, precise weapon drill was hampered and the supply of shot to the battlefield was unnecessarily complicated. By contrast, Wellesley ensured that his field gun was standardized and gun carriages and wheel contraptions were designed such that they were easily and efficiently hauled by bullocks in battlefields.

14. Quoted in Piers Brendon, *The Decline and Fall of the British Empire, 1781–1997* (London: Vintage, 2008), p. 199.
15. Morris, *Why the West Rules — For Now*, p. 509.
16. Paul Kennedy, *The Rise and Fall of the Great Powers: Economic Change and Military Conflict from 1500 to 2000* (London: Fontana Press, 1988), p. 191. Kennedy highlighted several economic and geopolitical trends of the post-1815 era for the rapid expansion of empires carved out by the Western powers, namely, the growth of international economy, the relative stability of Europe, the modernization of military and naval; technology over time, and the occurrence of merely localized and short-term wars.
17. Ibid., pp. 191–92.
18. Headrick, *Tentacles of Progress*, pp. 9–16.
19. Ibid., p. 16.
20. Robert Kubicek, "British Expansion, Empire, and Technological Change", in *The Oxford History of the British Empire; The Nineteenth Century*, edited by Andrew Porter (New York and Oxford: Oxford University Press, 1999), Chapter 12, pp. 248–49.
21. Richard Kesner, *Economic Control and Colonial Development: Crown Colony Financial Management in the Age of Joseph Chamberlain* (Oxford: Clio Press, 1981), p. 222. Kesner provides statistical data on colonial revenues versus expenditures, on trade, and on loans during the 1880–1914 period for all of the Crown Colonies, including the Straits Settlements. Chamberlain's efforts in developing the "large estates" of the British Empire did create new opportunities for colonial borrowing and did improve transportation communications. Despite this, Kesner maintains that Britain remained parsimonious with development funds and colonies were supposed to pay their own way and borrow only if revenue would increase to repay the loans.
22. M. Edelstein, "Foreign Investment and Empire, 1860–1914", in *The Economic History of Britain since 1700*, edited by R. Floud and D. McCloskey (Cambridge: Cambridge University Press, 1981) Vol. 2, p. 70.
23. Headrick, *Tentacles of Progress*, p. 14.
24. B.R. Tomlinson, "Economics and Empire: The Periphery and the Imperial Economy", in *The Oxford History of the British Empire*, edited by Porter, Chapter 3, p. 61.
25. For a recent study about Japan's (and China's) socialization into the European-dominated international order in the late nineteenth century, see

Shogo Suzuki, *Civilization and Empire: China and Japan's Encounter with European International Society* (London and New York: Routledge, 2009).
26. Richard Rosecrance, *The Rise of the Trading State: Commerce and Conquest in the Modern World* (New York: Basic Books, 1986).
27. Peter Burroughs, "Imperial Institutions and the Government of Empire", in *The Oxford History of the British Empire*, edited by Porter, Chapter 9, p. 174.
28. See E.H. Green, "The Political Economy of Empire, 1880–1014", in ibid., Chapter 16.
29. James L. Garvin, *The Life of Joseph Chamberlain* (London: Macmillan and Company, 1934), pp. 19–20.
30. Kesner, *Economic Control and Colonial Development*, Table 10, p. 86.
31. M. Haviden and D. Meredith, *Colonialism and Development: Britain and its Tropical Colonies, 1850–1960* (London: Routledge, 1993), p. 108.
32. Michael Adas, *Machines as the Measure of Men: Science, Technology, and Ideologies of Western Dominance* (Ithaca and London: Cornell University Press, 1989), p. 200.
33. Stephen Howe, "Empire and Ideology", in *The British Empire: Themes and Perspectives*, edited by Sarah Stockwell (Oxford: Blackwell Publishing, 2008), p. 166.
34. Ibid., p. 169.
35. Alison Twells, *The Civilising Mission and the English Middle Class, 1792–1850* (New York: Palgrave Macmillan, 2009), p. 219.
36. Sheldon Watts, *Epidemics and History: Disease, Power and Imperialism* (New Haven and London: Yale University Press, 1997), p. 43.
37. The missionary work and explorations of David Livingstone in Africa is often regarded as one of the greatest geographical explorations of the modern era and a testimony of the inhospitable environment of the "Dark Continent". As both a preacher and a doctor, Livingstone was ideally suited to the task of spreading Christianity and civilization together — although in 1848 he ceased his missionary work and became an accomplished explorer because he recognized that the wilderness itself had to be made more receptive to British civilization before any conversion of faith could take place.
38. Adas, *Machines as Measure of Men*, p. 9.
39. George Young, *Victorian England: Portrait of an Age* (Oxford: Oxford University Press, 1964), p. 6.
40. Quoted in *The Times*, 27 July 1925.
41. Quoted in Antony Thomas, *Rhodes* (London: BBC Books, 1996), p. 8.
42. See Michael Adas, *Dominance by Design: Technological Imperatives and America's Civilizing Mission* (Cambridge and Massachusetts: The Belknap Press, 2006).
43. Ferguson, *Empire*, p. 35.

44. See Zaheer Baber, *The Science of Empire: Scientific Knowledge, Civilization, and Colonial Rule in India* (New York: State University of New York Press, 1996); Roy MacLeod and Deepak Kumar, eds., *Technology and the Raj: Western Technology and Technical Transfers to India 1700–1947* (New Delhi and London: Sage, 1995).
45. Zaheer, *Science of Empire*, p. 237.
46. Ibid., p. 10.
47. Benjamin Kidd, *Social Evolution* (New York and London: Macmillan and Co., 1894), p. 318.
48. See Landes, *The Wealth and Poverty of Nations*, pp. 225–30.
49. Ibid.
50. Ibid., p. 432.
51. Clark, *A Farewell to Alms*, Chapter 16. Clark maintains that culture and social evolution, not institutions, shape economies. Despite the accessibility to new technologies and management knowledge, many former colonies of European empires were (and are) not able to close the income and growth gaps with the industrialized nations after 1945.
52. Ian Inkster, *Science and Technology in History: An Approach to Industrial Development* (London: Macmillan, 1991), p. 226.
53. Ferguson, *Empire*, pp. 368 and 371.
54. Niall Ferguson, *Civilization: The West and the Rest* (London: Allen Lane, 2011).

2
Pioneers of Change
Entrepreneurs and Engineers

The Industrial Revolution in the West introduced to the world a series of technological changes embodied in the development of railways, steamships, the telegraph, and the mechanized factory. They were made possible mainly by three categories of individuals — inventors, engineers and entrepreneurs. Inventors and engineers created things that have never been created before. Entrepreneurs think of ways of adapting the inventions to the needs of the consumer markets, that is, they turn an invention into an innovation. Thomas Edison, for example, is an inventor — but he was also an entrepreneur because he invented a feasible electric bulb, then designed a system that would deliver electricity to lighting customers, and eventually incorporated a number of companies to manufacture all parts and to supply the electric service. Colonial Singapore did not possess inventors who created products that were developed by entrepreneurs for the mass markets, or structures and projects that were completed by the skills and knowledge of locally trained engineers. It was strictly a busy trading outpost where the engine of commercial growth was driven by thousands of migrants from India, China and the Malay Archipelago, and parts of the British Empire. They were convinced that, with the protection of the British law, there were many pots of gold at the end of the rainbow. This chapter looks into the pioneering roles of nineteenth century individuals in laying the foundation of Singapore's entrepôt economy.

ROLE OF THE EUROPEAN AGENCY HOUSE AND THE CHINESE COMPRADOR

Beginning with the arrival in 1819 of Alexander Johnston who started the first European agency house in the following year, a steady stream of European merchants and entrepreneurs soon arrived and established agency houses and business operations, many of which were the forerunners of several big companies in Singapore today. Notable among them were Alexander Guthrie, Edward Boustead, William Paterson, Benjamin Keasberry, Abraham Logan, Robin Woods, and John Cameron.[1] As Singapore's trade grew the number of European firms also increased steadily, from 14 in 1827, to 36 in 1855, and 62 in 1872.[2] The number of residents in the European community, however, remained small and grew slowly, from 74 in 1824, to 360 in 1849, and 466 in 1860.[3] It was this small body of European merchants who provided an impetus to Singapore's growth. John Crawfurd, the Resident of Singapore, described the small European community in Singapore as "the life and spirit of the settlement" and insisted that without them "neither capital, enterprise, activity, confidence or order" would exist in the small colony.[4] John Thomson, the government surveyor, added that the Europeans were "principally following mercantile pursuits, and as a body, they were upright, honourable, and stable".[5] For the small but growing number of British emigrants to Singapore during the nineteenth century and the early decades of the twentieth century, coming to this part of the world was an experience by itself. Sailing in the passenger ships of the Peninsula & Orient, the Blue Funnel, the Glen Kine and the Bibby Line, many were from the middle classes back in Britain and a "career in the East, whether in government or business, offered a standard of living that could not always be guaranteed at home".[6] Although, as pointed out by Bernard Porter, there was a relative lack of an imperial education in their schools in the first three quarter of the nineteenth century, this did not make them ignorant of the empire created by their fellow statesmen.[7] Some corresponded with friends who had made the journey to the colonies. Long letters were received from middle-class colonists rambling about life in the settlement colonies. The newspapers and journals (such as the *Fortnightly*, the *Quarterly* and the *Illustrated London News*) also carried accounts of events in the periphery of the Empire. There were also popular shows and plays depicting significant events which happened in the colonies and exhibitions in London where colonial galleries were set up.

While there were prominent professional individuals who contributed in one way or another to the economic and social development of the island, the main agents of Western enterprise were the merchant and managing-agency firms.[8] The managing-agency system was designed to reduce the risks of foreign investment in underdeveloped countries where business expertise and management skills were scarce.[9] These expatriate firms had board of directors consisting overwhelmingly of British nationals. They maintained a close link with their counterparts in the large industrial cities of Britain, notably with Liverpool, Glasgow and Manchester throughout their years in Singapore. As Singapore's entrepôt trade grew, many of these agency houses also diversified their interests, such as holding agencies for banking and shipping lines, managing estate plantations, and later developing the tin mining industry. Founded in 1856, the Borneo Company Limited had been closely connected with tin mining in Malaya from the early years of its formation and imported modern tin dredging machinery for Lode Tin Mines of the Pahang Consolidated Company Limited, the Ipoh Tin Dredging and Temoh Tin Dredging. The company and its subsidiaries were also involved in the manufacturing of bricks and sanitary pipes, servicing of motor cars and in building construction and engineering. Similarly, merchant house Guthrie & Company Limited, founded by the Scotsman Alexander Guthrie in 1820, also had a long-standing connection to the tin mining industry. It bought and exported the metal from the Chinese tin miners until the coming of the European miner, with his modern appliance and methods. In keeping with changing tin mining technology, Guthrie — through its business associate Renong Tin Dredging Company — operated the first land dredger in this part of the world. Although primarily a trading company, Guthrie possessed a "selling organization staffed with technical and specially trained men, capable of handling the high grade products for which modern conditions create an ever-increasing demand".[10] Other well-known agencies which imported engineering and estate supplies, modern machinery and manufactured products were Harper, Gilfillan & Company, Edward Boustead & Company, and Henry Waugh & Company.

The European enterprises operated within a congenial political and administrative environment. Unlike the centrally planned economy of the Dutch East Indies since the introduction of the Culture System in 1830, British officials in London were not a vigorous force behind British economic expansion in Asia.[11] Similarly, British colonial officials did not engage in the actual conduct of productive and commercial enterprises.[12] Conversely, because of the early predominance of trading concerns, the Europeans in Singapore were quick to impose their collective interest on the authority. In February 1837, in response to a proposal by Edward

Boustead, the Singapore Chamber of Commerce was established "for the purpose of watching over the commercial interests of the Settlement" and that "all merchants, agents, ship-owners and other interested in the trade of the place" were eligible for membership.[13] In its early years the Chamber of Commerce agitated specifically on commercial matters, such as the suggestion to impose port dues in 1837 and for an improvement in steamship communication in 1845. Gradually, it began to assume a more political role but internal dissension and commercial rivalry between its members prevented the Chamber from achieving any success.[14] It needs to be stressed that, despite the presence of the European commercial class which dominated the colonial economy, they formed only part of a dynamic business community. In particular, Chinese merchants and those from the trading ports in Asia played a critical role in Singapore's growth as a regional trading centre, where exchanges between the local suppliers of raw materials and the manufacturers in Europe took place. Local merchants, however, were not able to rival the European counterparts because of two reasons. First, in an international trade where Britain and other Western countries supplied the manufactured goods and Southeast Asia essentially provided the raw materials and minerals, the European merchants were regarded as the key linkage of this exchange because they controlled the import and export sector through their agency houses. Second, the Europeans were capable of transferring needed capital for their commercial activity and skills for the extractive industries.[15]

The history of the Chinese as immigrants and settlers in Singapore has been well researched by scholars.[16] The Chinese knew the island of Singapore as "The Dragon's Teeth" since the fourteenth century. They had moved in and out of Singapore for a long time and a handful had remained on the island as cultivators. But circumstances changed when in 1819 Singapore was declared a free port by Raffles. There seems little doubt that the Chinese, especially the merchants, found foreign-dominated ports attractive so long as they were assured of safety from banditry and civil wars for themselves and their goods and also an authority which respect private possession of wealth and property. After the aftermath of the Anglo-Chinese War of 1839–42 and the Taiping Rebellion in the 1850s and 1860s, the *sinkehs* or "new arrivals", mainly from the provinces of South China, flocked to Singapore en masse.[17] Between 1840 and 1860 about 30,000 Chinese arrived in Singapore.[18]

This influx of people was what Singapore needed. The Chinese met the urgent demand for cheap labour to develop the town's infrastructure, open up the interior and, most important of all, work at the Singapore River and later at the New Harbour as coolies. They were also considered an invaluable source of entrepreneurial talent. In fact, so great was their number in Singapore that, from the late nineteenth century, British interests were involved in exporting indentured labourers to other underdeveloped areas of Asia. This became known as the infamous "Coolie Trade".[19] Interestingly, the diffusion of steam shipping and the establishment of regular steamship service in 1884 led to a strengthening of trade relations between Singapore and the British colonies in Western Australia.[20] The business of recruitment and transportation of Chinese coolies developed significantly and firms like Behn Meyer and McAlister and Company became the main agents of this trade. A majority of the Chinese coolies was channelled into the expanding pearling industry in Western Australia. In the process, Singapore also gained the reputation as a main labour transit centre for countries in Southeast Asia and the Pacific.

The Chinese in Singapore — and in Hong Kong — served two economic functions. For the merchants with entrepreneurial acumen, they imported the comprador system from China and acted as middlemen between the European agency houses and the inhabitants of the island. Hui Po-Keung explains this process in Hong Kong: "In the process of colonial expansion, Chinese businessmen played the role of middlemen. They served as intermediaries between European importers/exporters and native consumers/producers. This was due to the fact that the Chinese traders, lacking political support from China, in turn relied on the Europeans to explore and protect new business opportunities".[21] For the many unskilled immigrants, they provided the much-needed manual labour for the port, public works, and the rubber and tin industries. Thomas Braddell, a British resident, gave an insight to the workings of the Chinese comprador system in Singapore. Writing in 1855 in the *Logan's Journal* he said:

> The details of the great European trade of these settlements are managed almost exclusively by Chinese. The character and general habits of an European gentleman quite preclude him from dealing directly with the native traders, who visit our ports and bring the produce of their several countries.... Here the Chinese step in as a middle class and conduct the business, apparently on their own account but really as a mere go-between. The Chinese put himself on a level with the native traders, takes them to his shop ... and succeeds in dealing with the native on terms far inferior to what could have been obtained from the European merchants.[22]

A description of comprador Tan Kim Tian was provided by W.G. Gulland, an employee of the agency house Paterson Simons & Co and described by Trocki as "a pillar of the European mercantile community":[23]

> Mr Tan Kim Tian was compradore or head storekeeper to Paterson Simons & Co., which firm he entered as a small boy. He taught himself to read and write English and rose by degrees to the above position.... Chinaman like, he was many-sided and in addition to filling the above position carried on, on his own account, the business of merchant and steamship owner, having been one of the first in the coasting trade to replace sailing vessels by steamers.[24]

From the above source, two main characteristics of the Chinese compradors in Singapore could be seen. First, given the opportunity, he was likely to pick up a working knowledge of a Western language to facilitate doing business with the Europeans. Second, he was bonded, contractually and physically, to the European agency house over a long period of time. In the process, the "many-sided" comprador developed skills and assets which allowed him to expand his own business interests. If there was one reason why he did not give up his middleman role it was because he recognized the effective and profitable business arrangements which he had entered upon and which also offered better legal protection for his own personal advantage. Perhaps, this was one indication of the ability of the Chinese to adapt to change.[25] Indeed, the activities of the European and the Chinese traders were frequently complementary to the extent that Singapore's economic growth could be considered as a product not of British business but of Sino-Western enterprise.

While the Chinese merchant reaped the economic benefits of Singapore's growth, his counterpart, the thousands of unskilled immigrants eked out a hard life. But their "indomitable perseverance" received praise as far away as in Australia.[26] John Chinaman, as he was nicknamed by the Europeans, "is willing and able to perform those inferior offices for the European and other residents of Singapore at which other portions of the population would turn up their nose in supercilious disgust".[27] Besides their active role in the port, the Chinese were also involved in the economy as petty merchants and artisans. The former usually operated in small shophouses and retailed the daily needs of the local Chinese population, such as rice, spices and clothings. The artisans were mainly carpenters, builders of houses, bridges, boats and ships, blacksmiths, and so on. Unlike the powerful craft guilds in Victorian England, these artisans

Pioneers of Change

in Singapore largely operated independently and their fortune was more or less determined by the expanding entrepôt economy.[28] The myriads of occupations of the Chinese in Singapore, and certainly in most parts of Southeast Asia where they have settled, was provided by Seah Eu Chin in the *Logan's Journal* in 1848:

> The different trades and professions are school masters, writers, cashiers, shopkeepers, apothecaries, coffin makers, grocers, goldsmiths, silversmiths, tinsmiths, dyers, tailors, barbers, shoemakers, basket makers, fishermen, sawyers, boat builders, cabinet makers, architects, masons, lime and brick burners, sailors, ferrymen, sago manufacturers, distillers of spirits, cultivators and manufacturers of gambier and sugar, cultivators of pepper and nutmegs, vendors of cake and fruits, porters, play-actors, fortune-tellers, idle vagabonds - who have no work and of whom there are not a few — beggars, and, nightly, there are those villians, and thieves.[29]

Such was the depth of the role of the Chinese that they represented a powerful and influential force in shaping the development of the port and the economy of Singapore.

While the contributions of the Chinese coolie to the colonial economy are well documented, the role of convicts as pioneers of change in the transformation of the Singapore town is often overlooked. Right from the days of Raffles when they were used in the laying of roads, convicts had been widely deployed in public works throughout the nineteenth century. Major James Low, a prolific contributor to the *Singapore Free Press*, wrote in 1840:

> Singapore, Malacca, Penang and Moulmein are the Sydney convict settlements of India. There are upon an average about 1,100 to 1,200 native convicts from India constantly at Singapore. As at 1 May 1859, there were 2,330 convicts stationed in Singapore, the majority of whom came from Bengal and Madras. These are employed in making roads, assembly bridges and digging canals; and undoubtedly without them the town, as far as comfort in locomotion is concerned, would have been now but a sorry residence.[30]

It was noted by the Department of Public Works that, until the call for an extensive fortification of the colony, the convicts were mainly used in the "erection of new buildings and for keeping the old ones in repair".[31] Their skills were put into good use in the construction of the colony's fortification. Arriving in Singapore in January 1858, Captain Collyer of the Madras Engineers was appointed Chief Engineer and entrusted with the proposed plans for the fortification of Singapore, mainly on the South Battery at Government Hill (Fort Canning) and Fort Fullerton. He had at his disposal the whole labour of the convict body. It was to the credit of these convicts that the work produced was "considered to be of excellent quality, as good, if not probably better, than would be obtained from free Chinese labour; and so far the convict body has proved most useful in the new scheme of covering the hills and shores of Singapore with Batteries, Redoubts, Barracks, Magazines, etc".[32] Many of the convicts were riveters and blasters. In addition, the convicts were also skillful at brick-making and put to good use. Captain McNair of the Madras Artillery and Chief Engineer of the Convict Department stated in his report that "so highly are convict-made bricks appreciated in the place, that in some private establishments every endeavour is made to secure the services of convicts … to assist in the manufacture [and] the knowledge convicts have also of the mode of keeping every roads in order, the result of long training in the construction of nearly every road in the Island, is taken advantage of…"[33]

EARLY ARRIVAL OF THE BRITISH ENGINEER

While the contributions of the European and Chinese entrepreneurs to Singapore's growth were immense, the role of the British engineers was equally significant. The British engineers of the early decades of the Industrial Revolution were largely men from humble backgrounds, such as artisans, mechanics, millwrights, instrument makers, and stonemasons, with limited formal education. The passing of the eighteenth century in Britain, however, marked the gradual displacement of the millwright, a skilled worker in wood and iron, by the new engineer, a creative professional skilled in the fusion of engineering and the industrial arts.[34] Technically trained people began to play a role in shaping the way people lived and identifying themselves as members of the new engineering profession. Many of them were probably involved with driving steam locomotives and tending stationary steam engines and (what we would now designate as civil engineering) building canals, surveying roads, constructing railway lines, dock construction and shipbuilding, and supervising construction crews. Few of these engineers in the pioneering days had had any formal education either in the sciences or in what were then called the technical arts. They learned their craft working on the canals and railroads and as apprentices in machine shops. On-the-job training could have a taught a young person to practise engineering because most schools, in the nineteenth century, did not teach anything vaguely resembling science, let alone engineering.

The growth of the professional engineer in Britain was remarkable. The total number was less than 1,000 in 1850; in 1870 it had risen to 4,128, to 15,043 by 1890 and to 40,375 by 1914.[35] In the United States, the 1850 census was the first one to include "engineer" as an occupation and about 2,000 people were included in the survey.[36] With the birth of the steam fleet, engineers were recruited in large numbers as officers in the British Royal Navy, to operate and maintain the engines.[37] In the hands of these professional individuals, Britain's commanding world position of the Victorian era (1837–1901) was built. They also played no small part in projecting technology as a powerful tool of imperialism by crossing the oceans to impart their knowledge and skills to indigenous peoples under the protection of the British Crown. The arrival of British engineers into tropical colonies like Singapore was part of a larger movement abroad of these skilled professionals from Britain.[38] The British engineer became an agent of Britain's "engineering imperialism" and for the transfer and diffusion of technology into the many colonies in Asia and Africa. In the words of Buchanan:

> the great expansion of British industrial activity in the 19th century, engineers shared in the general entrepreneurial euphoria and were anxious to pursue any opportunities for the acquisition of fame and fortune…. the transfer of Western expertise to the comparatively underdeveloped parts of the world involved a more enduring contribution by British engineers…. The impact of the diaspora was thus greatest in the British empire, but its influence survived the dismemberment of the empire and equipped the new nations that emerged from it with a sound infrastructure of transport systems and urban services.[39]

British engineers found their way into Egypt, Central and Southern Africa, India, Malaya, Singapore, China, Japan, Australia, and elsewhere. In some larger countries like India and Australia these engineers contributed to a "systematic colonization" of lands through their various engineering schemes.

There is a dearth of source material pertaining to the emigration of British engineers into Singapore and Malaya, especially during the nineteenth century. Presumably, the number of these professionals must be few. One source listed the names of forty-one European engineers living in Singapore in 1904 and who also served as qualified jurors.[40] Those who did find their way to Singapore were mainly linked to the shipbuilding industry. Singapore's reputation as a thriving port was a sufficient pull factor. Nevertheless, by the end of the nineteenth century, there were several qualified engineers in Singapore who were mainly employed with the Tanjong Pagar Dock Company and engineering firms, Riley Hargreaves & Company and Howarth Erskine Ltd. The first trickle of the arrival of British expertise took place not long after 1819. They were mostly artisans, mechanics, shipwrights, and military engineers belonging to the East India Company. Two such individuals who contributed in no small measure to the development of Singapore during the early decades of the nineteenth century were architect and civil engineer George Drumgoole Coleman and the surveyor John Turnbull Thomson.

In the age of Imperialism, colonial officials in British India and the Straits Settlements fully promoted the construction of European-style housing and official buildings because they reflected their sense of power and prestige. Singapore was no exception, especially when it was gradually seen as a symbol of Britain's economic prowess in Southeast Asia. In 1833 George Drumgoole Coleman came to Singapore after spending several years in Calcutta and Batavia. He was a trained civil architect at a time when colonial buildings were usually erected by a handful of military engineers, surveyors, and draughtsman. He was appointed as the first government Superintendent of Convicts and Public Works and was responsible for planning the centre of the Singapore town. In the hands of Coleman the Palladian style of Georgian England and of Colonial India was transferred into the growing urban landscape of the Singapore town.[41] Many today, however, would argue that Coleman boldly copied his building designs from his extensive references to printed works of architectural designs. He cleverly combined the characteristic styles of the Western architecture with the practical requirements of the tropics. As a result, he developed the use of wide verandahs for shade and louvred windows to reduce glare and provide protection against rain and yet allow thorough ventilation.[42] The verandah and louvred windows still provide a sense of colonial aesthetic to some modern homes in Singapore today.

Singapore's early development also owed much to the Scottish surveyor, engineer, and architect, John Turnbull Thomson, who arrived in Singapore in July 1838.[43] Three years later he was appointed government surveyor and remained in office for a total of twelve years. Like Coleman, Thomson constructed a number of buildings, bridges, and roads. He was the first to introduce and use "Portland cement" — a type of hydraulic cement invented in England in 1824 — to enhance the quality of construction.[44] Thomson also made several elaborate surveys of the Straits of Malacca, the east coast of the Malay Peninsula and the New Harbour in Singapore and many of these were compiled and used by the Admiralty.[45] But Thomson's place in the history of Singapore's technological development is in his construction of the Horsburgh Lighthouse in Singapore Straits (54 kilometres from the harbour) in 1850.[46]

The lighthouse was seen by the European powers as a significant tool of empire building. Lighthouses belonging to the various European powers, particularly the Dutch and the British, sprouted along the sea routes to the East, stimulated by the expansion of world trade with the opening of the Suez Canal in 1869 and technological advancement in steamship construction. By the 1880s, lighthouse technology too had advanced rapidly compared to the 1820s and 1830s. Engineers were able to develop a bright light magnified by lenses of just the right design and used group flashing method in the case of revolving lenses. According to an 1886 publication by the Royal Geographical Society, 171 lighthouses, 72 of which were along the Indian sub-continent and in Sumatra, were erected by Britain to protect its lucrative trade routes to the East.[47] The idea for a lighthouse in the Singapore Straits was first mooted on 2 November 1836 at a meeting in Canton, China, when British merchants voted to raise funds for a lighthouse to be named after the navigator and hydrographer Captain James Horsburgh. The decision was also hastened by the fact that between 1822 and 1839, five ships had been wrecked, and in the ten years before the lighthouse was commissioned a further ten ships were lost. The site chosen was the island of Pedra Branca. About the same time, a similar decision was made by merchants plying their trade in the Indian Ocean to erect the light at Cape L'Aghulas in the Cape Colony. Both initiatives illustrated how local interests and private enterprises were more responsive to business needs than the authorities in London.

Construction of the Horsburgh Lighthouse started in May 1850 and on 15 October the light was lit for the first time. It was an engineering feat which combined Western technology and the ingenuity of the Chinese artisans, mainly stone cutters and masons, and carried out in considerable difficulties. Huge pieces of granite from Pulau Ubin (a small island situated northeast of Singapore) were cut into 5,474 cubic feet (155 cubic metre) blocks and transported to Pedra Branca or "White Rock". Here Chinese coolies, together with Malay workers and Indian convicts, hoisted the stones, some weighing as much as seven tons, by ropes and cross-stretchers. Their task was to erect the first lighthouse in this part of the world and the first built in granite masonry. Besides granite blocks, local timber was used and brass rails and stair-case were mould in Singapore. Only the cast-iron dome and lanterns were imported from Messrs. Stevenson of Edinburgh.[48] Built as a cost of 53,020 rupees, the 109 feet tall Horsburgh Lighthouse with seven levels of rooms, was eventually completed in September 1851 and went into full operation.[49] The lighthouse soon took on the same mythical reputation as Bell Rock and Skerryvore, becoming the "First Pharos of the Eastern Seas" or "Lighthouse for All Nations". Its revolving light "was easily visible at 14 miles, and its position is a very admirable one".[50] It was a fine engineering achievement, reflecting Thomson's own inner conviction and drive. In the race to illuminate the world, "the winner in South-East Asia was the wealth of all nations, but particularly Dutch and British local economic interests".[51]

The scarcity of capable engineers in Singapore at this time was matched by a low level of technical and surveying skills on the part of military and civil engineers stationed in Singapore. A good example was the engineering effort of one Captain Faber of the Madras Engineers. In September 1843 Major Faber, the newly appointed Superintending Engineer, was called on to build a bridge over the busy Singapore River. Unfortunately, he built the bridge so low that the lighters could not pass under it at high tide, and he was amazed at the negative response to his suggestion that the river under the bridge should be dredged to lower the water level.[52] Faber's misfortune continued when the roof of the landing platform which he built on the river collapsed because "the pillars were too thin".[53] Faber even fancied himself as a qualified architect and built the Local Government Offices which was so poorly designed and constructed that part of its roof collapsed in 1846 and injured a number of people. It was discovered later that the pillars supporting the ceiling were far too thin. The "Faber episode" was aptly described by Benjamin Cook in one of his letters:

Pioneers of Change

The Major arrived from India in the early Forties as Superintending Engineer for the Company. An early assignment was the New Market whose walls cracked dangerously and had to be rebuilt, then a roof he was constructing at Boat Quay fell down and killed some workers, and the flag-staff on the grandly named Mount Faber, being inadequately conducted, was within a month struck by lightning and destroyed. The Grand Jury then complained that two little bridges he had constructed across the Canal were, by their flatness, obstructing boats at high tide so that their centres should be raised. But the Major, no doubt believing that even the sea would bow to the demands of a servant of the Honourble Company, said that instead he would dredge the Canal to lower the water level![54]

In another early engineering project, a proposal was made to improve on the defence of Singapore in 1859. Government Hill where Raffles first built his residence was seen as the highest and most strategic point to defend the seaward approach to the port. Accordingly, it was levelled off, seven 68-pounders were installed and the hill renamed as Fort Canning. However, when the task was completed, "it was noticed that Pearl's Hill was higher; so the military engineers had to cut down the top of that hill!"[55]

These slip-ups could be due to lack of hindsight or some miscalculations on the part of the military engineers but they did imply a general low level of technical proficiency. It is also interesting to note that completed frameworks for the colony's public infrastructure, such as the Elgin iron bridge in 1863, were imported all the way from Britain. In most cases, however, bridges were locally constructed and required constant upgrading due to poor materials and workmanship. It was only in the 1880s that iron, rather than wood, was the main material used in bridges and culverts because the "savings in wear and up-keep would be very great, as the 'life' of a wooden bridge can hardly be reckoned at much more than five to six years, with a heavy cost for up-keep, while that of iron would probably be at least 20 years with a less cost for up-keep".[56] The quality of public works structure was also called into question. As reported by John Douglas, the Colonial Secretary of the Straits Settlements in 1877: "Fresh disappointment was in store for the Singapore Public respecting the waterworks, which were at one time confidently expected to be opened by the close of the year. But the collapse of one of the low level reservoirs, and the general failure of the foundation of the engine and boiler house again caused delay, necessitating the removal of the machinery and the underpinning of the foundations of the building".[57] Furthermore, despite the early existence of a survey department, by the 1880s, there was still "clearly the absence of any systematic survey or professional guidance" and that "the work had, in most cases, to be done over again".[58] By the first decade of the twentieth-century, the efficiency and quality of surveying and engineering works did not seem to improve, as indicated by a letter to the editor of the *Straits Times* in 1914 regarding the repair of the Grove Road Bridge:

> For the past eight months, to my knowledge, the bridge has been undergoing repairs and now, at this late date, it has actually been closed to all traffic. Last night, I was surprised to see that no night work was being done. Now, if the municipal engineer in charge of this work has any consideration for public feeling whatsoever, surely he would see that the work is carried out as expeditiously as possible, working night and day until the bridge can be reopened to traffic.... How long is the work of repairing to go on? Judging from past instances, when only about half a dozen of coolies have been employed, the work will go on for many moons yet and be a good source of income to the contractors at the public expense.[59]

Notwithstanding the general negative perception of the quality of engineering works, British engineers and architects did produce several outstanding projects which, till today, are earmarked as iconic buildings and ornamental structures in the urban landscape of Singapore. Swan and Maclaren were both surveyor-engineers and the firm they founded in 1885 became the most important and established architectural firm till the 1930s. The firm was unrivalled in terms of its standard of architectural work and this was largely credited to the entry of R.A. John Bidwell as a partner of the firm in 1900. Bidwell's experience was wide-ranging. He had served with the London County Council and worked in private firms. Bidwell was also a member of the Architectural Association in London. As an agent of change and regarded as one of the two (besides Coleman) truly professional architects to live in Singapore before the First World War, Bidwell was largely responsible for the design and construction of some of Singapore's existing buildings of historical significance — Raffles Hotel, the Victoria Memorial Hall and the Teutonia Club (now the Goodwood Hotel).[60]

CONTRIBUTIONS OF BRITISH PROFESSIONAL ENGINEERS

While the British engineer in Singapore did not construct monumental engineering works which his counterpart created in Egypt, India, Africa, and Australia, he was, nevertheless, instrumental in developing the shipbuilding industry, public works, mining and plantation ventures. In Britain, it was only towards the end of the nineteenth century that full-time engineering education at a tertiary level started to increase significantly.[61] In line with the increasing importance of engineering in Britain, more professional mechanical and civil engineers arrived in Singapore and Malaya and started to provide supportive engineering services.

The history of shipbuilding in Singapore dated back to 1829 when William Temperton, a shipwright, laid down the keel of the brig *Elizabeth*, the first sea-going vessel built along the banks of the Singapore River. In 1848, the site was acquired by Wilkinson, Tivendale and Company. Thomas Tivendale and James Baxter constructed Singapore's first steamer, *Ranee*, an anti-piracy gunboat fitted with an imported 4-horsepower engine from London. Two more gunboats, the *Malacca* and the *Singapore* were subsequently built with imported engines by the company in 1856. But, by this time, the technology utilized was considered outdated. The gunboats were built to "cruise after petty Pirates, but unless with a strong breeze in their favour, they are not calculated for pursuit".[62] This was also the case for the cruiser *Hooghly*, a 192 tons and 50-horsepower anti-piracy steamer used by the Marine Establishment. Her engine was "antiquated [and] as a cruiser after pirates she is inefficient, owing to want of speed and to heavy draft of water".[63]

The quality of ships built improved tremendously after 1870 when a number of professional engineering firms were set up in Singapore. In 1871 Buyers and Robbs launched the *Bintang*, a forty-ton steamer, the engines of which were built locally by Riley, Hargreaves and Company.[64] Richard Riley came to Singapore as a civil engineer in 1868, and together with shipwright William Hargreaves, established the engineering firm. It had a large engine shop, machine shop and building sheds, well equipped with powerful punching and shearing machines, bending machines, plate rollers, and bar furnaces.[65] It was fully adapted for shipbuilding in steel, iron, composite, or wood. The company employed several qualified engineers in 1904 — John McLachlan, Herbert Saxelby, James Allan,

James Armstrong, Win Chalmers, James Drysdale, Robert Goldie and Samuel Smith. Two other prominent engineering firms, Howarth Erskine Limited and Central Engine Works, also provided a range of engineering services. Howarth Erskine Limited also had several professional engineers — William Anderson, Alexander Maclennan, Walter Palliser, James Quinn and William Sharp. The two well-established engineering firms amalgamated in 1912 to form the United Engineers Limited, the largest engineering firm in Singapore till the 1950s. By 1929, the company employed almost 2,000 workers and had five branches in Malaya.

The Tanjong Pagar Dock Company was another strong competitor in the shipbuilding industry. It was seen as the largest industrial enterprise in the colony at this time.[66] To meet the worldwide expansion of shipping activities, construction of wharf facilities was mainly carried out by the Tanjong Pagar Dock Company. Between 1869 and 1874, the company engaged 22 British engineers to man the whole operations of the company. They were assisted by 350 carpenters and 100 blacksmiths and boilermakers and hundreds of unskilled dock workers.[67] Using its own facilities and expertise, the Tanjong Pagar Dock Company built the largest of their vessels, the 448 tons *Bentong* in 1903. In that year, too, Riley, Hargreaves and Company launched the *Sea Mew*, a government steamer. The efforts of the British engineers and the quality of the ships built prompted Frank Swettenham, the Governor of the Straits Settlements, to endorse the policy that "it must be borne in mind that work for the Straits should be given in the Straits, if costs and workmanships are equal".[68] Besides handling the technical aspects of their jobs, British engineers in the shipping industry, such as, Richard Black, Corlis Cancon, John Gartshire, Robert Lindsay, David Munro, John Murray, Cyril Neubronner, Frederick Niblock, William Paxton, Alexander Sharp, John Sunner, John Tobias, and Nicol Weatherstone, had to train apprentices in repair and construction works. A range of workshop facilities were erected in order to meet with the more sophisticated steamships that called at the harbour.

To sustain their social and professional aspirations, and also to strengthen their sense of identity, the British engineers formed the Association of Engineers on 7 December 1881. Its interests were largely centred on the shipbuilding and repair industry. Thus, it aimed to "guard the interests, promote and further the welfare, elevate and improve the condition of all connected therewith by the diffusion of sound practical knowledge, [and] to give steamship owners greater facilities for obtaining sea-going engineers of undoubted practical experience and ability".[69] The office bearers and members were Scots from the few British engineering and surveying firms, forming "a merry crowd, mostly hailing from the Clyde".[70] It was an exclusive European institution. J.H. Drysdale, who arrived as an engineer in March 1872, affirmed that "the engineering community was ninety per cent Scotch".[71]

British skilled workers also played a part in starting industrial ventures in Singapore. In 1883 two enterprising printers, John Fraser and David Neave, decided to diversify their printing operations (The Singapore and Straits Printing Office at 100 Robinson Road) and so formed the Singapore and Straits Aerated Water Company. Both operations were well equipped with imported machinery from England. Their modest factory operations, based on the principle of division of labour, were upgraded with the arrival of A. Morrison, an engineer with considerable experience in aerated water manufacture.[72] In 1892 the company advertised in the local newspapers as "manufacturers of soda water, selter water, potass water, lemonade, tonic, ginger ale, ginger beer, etc. of the first quality only".[73] The company's local clients included the main clubs, hotels and private residences. They were also sold in Australia through appointed agents. Six years later, on 27 January 1898, the company advertised in the local newspaper to raise share capital of $225,000 and had its name changed to Fraser and Neave Limited. The company manufactured "all descriptions of aerated waters of first quality only" and its printing department was "very completely equipped with plant of the most modern description and being, like the AERATED WATER DEPARTMENT, under skilled practical management, the best class of work is guaranteed".[74] The factory was one of the earliest establishments to enjoy the wonders of electricity which "will, when completed, enable the outputs to be largely increased, besides conferring other advantages on the business".[75]

In Malaya, too, British engineers were creating a modern infrastructure by building miles of roads, railways and constructing bridges. According to the British administrator Frank Swettenham, railway construction was "carried out by engineers in the service of the States, from the first trial surveys to the public opening and working of the lines, without any assistance, except an occasional reference to the consulting engineers when advice was required on any matter of unusual importance, such as a design for a large bridge".[76] This practice was considered unusual since the established system was for the Crown Agents to appoint consulting engineers to undertake the task. The latter, in turn, recruited engineers and sent them out to the colonies. Nevertheless, the railways constructed in the Malay Peninsula were of such high standard that they were being described as "too good".[77] Swettenham gave the impression that there was a sizeable number of British expertise working on the Malayan railways and that they were largely responsible for bestowing the country with miles of railways. In her study of the railway building and its impact in Malaya during the years 1880 and 1957, Amarjit Kaur highlighted the significant role of the skilled and semi-skilled Indian workers who were recruited in thousands from British India to build and operate the system. Many of these workers had skills and experience in railway building.[78]

By 1940 Britain's scientific and technical individuals had transformed the once swampy settlement into one of the world's busiest ports and a modern city. As part of the diaspora of British engineers migrating into peripheral colonies of the British Empire during the late nineteenth and early twentieth centuries, these agents of change transferred Western science and technology into the trading emporium. Their contributions were all the more significant because of the absence of native experts. Sjovald Cunyngham-Brown, an engineer who served in the Malayan Civil Service in the 1930s summarized the contributions of the various departments, in particular, the Public Works Department (PWD):

> We would have been nothing had it not been for the police, the Education Services, the Health and Medical Services, the Survey Department from New Zealand, the Mines Department that discovered all the areas for private interests to exploit, the Agricultural Department — who were practically the originators of the wealth of Malaya — to say nothing of that most silent, unobtrusive and generally forgotten arm of government known as the Public Works Department ... with all their roads, their bridges and their brothers-in-arms, the Electrical Supply Department and the Wireless and Telegraph Services, that quietly brought a civilised country into being. There was nothing that the PWD did not do. They created all the furniture for the government offices and bungalows — and built the bungalows themselves.[79]

Despite the establishment of the PWD and the extensive availability of technical experts in private enterprises, the diffusion of technical knowledge and skills to the locals was at best minimal. Daniel Headrick has argued that the transfer of Britain's technology, including scientific knowledge

Pioneers of Change

and technical skills, into its imperial colonies like Singapore and Hong Kong was more geographic rather than cultural.[80] It was more a process of "relocation, from one area to another, of equipment and methods, along with the experts to operate them".[81] Skilled Westerners rarely settled in the tropics, but came for a few years, segregated themselves from the native society, then returned home. In Singapore, the majority of British engineers and skilled Europeans had their own social enclaves, and social interaction with local workers was minimal.

TECHNICAL KNOWLEDGE TRANSFER AND RACIAL PREJUDICE

Imperialists like Swettenham, Weld and Winstedt were upholders and practitioners of a conservative, traditional and ordered British Empire based not exclusively about race or colour, but also about class and status. In his analysis of how the British saw their own empire, David Cannadine stated that the British throughout the Empire were "tied together by a shared sense of Britishness" [and] the phrase that best describes this remarkable transoceanic construct of substance and sentiment is *imperialism as ornamentalism*".[82] He concludes that the British people possessed a "hierarchical-cum-imperial mindset" which promoted the "cultivation and intensification of racial differences based on post-Enlightenment attitudes of white and western superiority and of coloured and colonial inferiority (along with the cultivation and intensification of gender differences based on attitudes of white and male superiority and white and female inferiority)".[83] In short, the Westerners' confidence in their own racial, scientific and technological superiority "has prompted disdain for African and Asian accomplishments, buttressed critiques of non-Western value systems and modes of organization, and legitimized efforts to demonstrate the innate superiority of the white 'race' over the black, red, brown, and yellow".[84] One contemporary British civil engineer, G.J. O'Grady, who served in Singapore from February 1928 to February 1932, commented at length on the issue of "colour bar":

> We came to these countries (The Malay Peninsula) by right of conquest and established ourselves there on that basis — just as Australians did in Australia and Americans did in America. There was one big difference. We did not in these cases destroy the conquered races. We let them live.

> Everyone of us in our school boy story books had read hair raising tales all of which end in the final triumph of the superior white man.... We have since those days of conquest changed the entire basis of our own social structure. We have become more tolerant.... There is, however, one colossal lag, which is not of our making. The social education and aptitude for full democratic rights among Asiatics are centuries behind us. This is compounded of superstition, religion, the Caste System, and above all THE LACK OF ANY IMPULSE TOWARDS ATTEMPTING TO ALTER THEIR AGE-OLD SOCIAL STRUCTURE THEMSELVES. Still we are human. When we relax, we like to relax in our own way, and some will inevitably go too far, with consequent loss of dignity. It is very hard to relax under the watching eyes of those of another race and creed and way of life. So we badly need the protection of the colour bar in our clubs and social hotels....[85]

The colonizers' usage of technological and scientific gauges of human potential has tremendous implications on the nature of Western policies regarding education and technological diffusion which serve to shed light on the uneven levels of underdevelopment in the developing countries today. In some ways, such a situation did occur in colonial Singapore — through an "enclavist" European community and their monopoly of all important posts in the private and public sectors which required a certain level of scientific and technical training and experience.

As controller of the administration and economy of the Straits Settlements and Malaya, the British occupied high posts and enjoyed a relatively good life. As the anti-imperialist American scholar Rupert Emerson puts it, "to undertake manual labor, to live at something approaching the level of the general public, is in practical fact regarded as a betrayal of the white mission of superiority (and) it is not that there are some Europeans who are living luxuriously, but that, broadly speaking, there are none who do not live in that fashion".[86] Looking at the issue objectively, it cannot be denied that official policies adopted in dependent colonies of the British Empire reflected a general segregationist approach. The British sought to develop their own exclusive community even though within itself some form of social distinctions between the various groups of Europeans existed. While praising the Chinese as the "bone and sinew" of the economy, Frank Swettenham added that it "is almost hopeless to expect to make friends with a Chinaman, and it is, for a government official, an object that is not very desirable to attain".[87]

Clearly, Swettenham was giving a warning that in order to maintain their superior position, a gap between the Europeans and the non-Western local population must be preserved; treating a Chinese as a friend would mean a closing of the gap.[88] In a similar vein, Frederick Weld famously declared: "I doubt if Asiatics will ever learn to govern themselves; it is contrary to the genius of their race, of their history, of their religious system, that they should. Their desire is a mild, just and firm despotism: *that* we can give them".[89] This "official mind" was pervasive and it is not uncommon for young British civil servants who were considered for appointments in the Chinese Protectorate to deliberately fail the test in Chinese language.[90] This distanced relationship with the locals was also testified by several British personnel working in Singapore and Malaya who maintained that "outside working hours there was little contact with Asians".[91]

This social exclusivism was further perpetuated by the government, European companies and managing agencies recruitment of European expertise to handle all middle and high management, technical and engineering jobs.[92] As late as 1940, in government institutions like the Singapore Municipality, all engineers in the Gas, Water and Electricity Departments were recruited from Britain. A small number of Chinese worked as inspectors, assistant superintendents, assistant architects and assistant surveyors. In the case of the Singapore Harbour Board, Europeans were employed as Chief Engineer down to the machine shop foreman. The Chinese took up most of the clerical and cashier positions.[93] British engineers and technical staff also occupied most of the engineering establishments. This does not imply, however, that the supply of British engineers met up with the demand for their expertise in Singapore. In his personal account as a civil engineer in Malaya and Singapore, G.J. O'Grady mentioned that, during the late 1930s, development works by the Singapore Rural Board were frequently delayed because of a shortage of engineers.[94] The government also ensured that British personnel never suffered what was regarded as the indignity of serving under Asians. White engine drivers, bootblacks, prostitutes and others employed in menial occupations were deported. And, while non-Europeans could visit the famous Raffles Hotel, they were not permitted on the dance floor.[95] Even within the British working community, there existed some form of social segregation. A British newcomer "quickly learned to know his place, which in the British community was very largely decided 'by your work and for whom you worked' [and] in business circles this meant that a social barrier divided mercantile from trade, Europeans working in such large stores as John Little, Robinsons or Whiteaway and Laidlaws were held to be tradesman and were expected to stick within their own social circles and clubs, as were a handful of other salaried workers such as the British or Eurasian engine-drivers, NCOs on attachment from the British Army and jockeys".[96]

While no specific examples in Singapore are being mentioned here, there were development projects in Malaya which reflected the dependency on British engineers for the design and completion of engineering works. In a project to supply water to Port Dickson, "what is believed to be the largest order yet placed in Great Britain for weldless steel mains for service in Malaya has just been secured by the British Mannesmann Tube Company Limited".[97] The contractor was Paterson, Simon and Company Limited in Singapore and to complete the British connection "the whole of the material is being carried from England in British steamers".[98] In another development project, the Rapid Gravity Water Filter at Province Wellesley was built solely by British expertise.[99] The mechanical equipment was manufactured and supplied by Bell Brothers Limited in Manchester, through the Crown Agents for the Colonies, while the contract for the reinforced concrete construction work and installation of the mechanical equipment was carried out by the engineering department of Paterson, Simon and Company. British technical expertise was also used by Chinese industrialists in Singapore. Indeed, the British engineer seemed to possess a special halo because it even motivated one Chinese-owned engineering firm, Eureka Motor and General Engineering Works, in its advertisement in the *Singapore and Malayan Directory* for 1922 to highlight the statement "English Engineer in Charge".[100] The majority of engineers and skilled Europeans had their own social circles, and social interaction with local workers was minimal. In the British-controlled rubber industry, there was hardly any diffusion to the local population of the findings of scientific research on the rubber industry carried out by the government because most of the research done was for the benefit of the European-owned plantations.[101]

In short, the technical and engineering profession in Singapore during the colonial period was the prized possession of the British and other Europeans. Buchanan reiterated that "the social effects of British engineering were even more attenuated by the presence of a large native population without any permanent European settlement [and] large strategic installations at Singapore and Hong Kong encouraged trade and generated vigorous urbanization, which in turn led to a need for civic buildings and public works such as water supply and waste removal".[102] For an old Malayan who could look back some years, the greatest changes which had been wrought by the British architect and engineer were found in the suburbs of the Colony - where the British lived: "[T]he pleasant conditions of living which they had created for themselves in a land which was once, just malaria ridden swamps [and] it was hard, indeed, to be a citizen of Singapore or Malaya and not feel the intense triumphant pride of creation — pride of our skill in engineering, organisation, medical lore, development and in sum our triumphant civilising mission".[103] These words reflected the imperial attitudes of the highly educated and much travelled Englishmen who prided themselves as the giver of the triumphant civilizing mission. No other areas illustrated more vividly the pride and contributions of British technical and engineering expertise than in the construction and expansion of the Singapore port.

Notes

1. Unlike Alexander Guthrie, Edward Boustead and William Paterson who were established merchants, John Cameron, Abraham Logan and Benjamin Keasberry were in the business of newspaper production and printing enterprises. John Cameron was the proprietor-editor of the *Singapore Free Press* which started circulation in 1835, while Benjamin Keasberry started a mission printing press which was bought over by the Fraser and Neave partnership.
2. Wong Lin Ken, "The Trade of Singapore, 1819–1869", in *Journal of Malaysian Branch Royal Asiatic Society* XXXIII, no. 4 (1960): 167.
3. Mary Turnbull, "The European Mercantile Community in Singapore, 1819–1867", in *Journal of South East Asian History* X, no. 1 (1969): 13.
4. John Crawfurd, *Journey of an Embassy from the Governor-General to the Courts of Siam and Cochin China* (London: H. Colburn and R. Bentley, 1830; reprinted Kuala Lumpur, 1967), p. 383.
5. J.T. Thomson, *Some Glimpses into Life in the Far East* (London: Richardson & Co, 1965), p. 202.
6. Charles Allen, *Plain Tales from the British Empire* (London: Abacus, 2008), p. 476.
7. Bernard Porter, *The Absent-Minded Imperialists: Empire, Society, and Culture in Britain* (Oxford: Oxford University Press, 2004), p. 83.

8. G.C. Allen and A.G. Donnithorne, *Western Enterprises in Indonesia and Malaysia: A Study in Economic Development* (New York and London: Allen & Unwin, 1957), p. 52.
9. Ibid.
10. *British Malaya*, December 1937, p. 193.
11. R.P.T. Davenport-Hines and Geoffrey Jones, *British Business in Asia Since 1860* (Cambridge: Cambridge University Press, 1989), p. 24.
12. Ibid.
13. *Singapore Free Press*, 2 February 1837.
14. Mary Turnbull, *The Straits Settlements, 1826–67: Indian Presidency to Crown Colony* (London: Athlone Press, 1992), pp. 137–38.
15. Lee Poh Ping, *Chinese Society in Nineteenth Century Singapore* (Singapore: Oxford University Press, 1978), p. 6.
16. See Chan K.B. and Claire Chiang S.N., *Stepping Out: The Making of Chinese Entrepreneurs* (Singapore: Prentice Hall, 1994). It contains a comprehensive bibliography on the topic of Chinese migration and settlement in this part of the world.
17. Ibid., p. 168.
18. Lee, *Chinese Society in Singapore*, p. 38. As part of the Chinese diaspora escaping from the turbulent decades of the 1840s, 1850s and 1860s, more than 100,000 Chinese also came to the "Gold Mountains" of California during the 1849-era gold rush to make their fortunes. The story of the American Chinese is vividly told by the late Iris Chang. See Iris Chang, *The Chinese in America: A Narrative History* (New York; Penguin Books, 2003).
19. The Chinese word *ku-li* literally means "hard strength" and it was popularly used by foreigners living in China who employed Chinese as household helpers or menial labourers. The term takes on a different colouration in the 1840s, when European capitalists — collaborating with unscrupulous Chinese recruiters or scrimps — started to recruit Chinese men to work on colonial plantations in regions like South America and the Caribbean. It was estimated that by 1870s, an estimated three-quarters of a million Chinese men became victims of the coolie trade.
20. See Jan Ryan, "The Business of Chinese Immigration", in *Private Enterprise, Government and Society: Studies in Western Australian History*, edited by Frank Broeze (University of Western Australia: Centre for Western Australian History, 1992), pp. 24–35. Ryan's research is found in his "A Study of the Origin and Development of Chinese Immigration into Western Australia", unpublished Ph.D. dissertation, University of Western Australia, 1989.
21. Hui Po-Keung, "Comprador Politics and Middleman Capitalism", in *Hong Kong's History: State and Society Under Colonial Rule*, edited by Ngo Tak-

Wing (London and New York: Routledge, 1999), Chapter 3, pp. 36–37. Hui argues that explanation for Hong Kong's economic growth before 1940 is not solely attributed to the stable colonial rule and its free trade policy. The role of the compradors is also a critical explanation.
22. T. Braddell, "Notes on the Chinese in the Straits", in *Journal of the Indian Archipelago and Eastern Asia* IX (1855): 109–24. J.R. Logan was the editor of this journal which accumulated many invaluable contemporary articles over the years 1847 to 1859. Hence, it was popularly known as *Logan's Journal*.
23. C.A. Trocki, *Opium and Empire: Chinese Society in Colonial Singapore, 1800–1910* (Ithaca and London: Cornell University Press, 1990), p. 52.
24. Quoted in Song Ong Siang, *One Hundred Years' History of the Chinese in Singapore* (London: John Murray, reprinted, Singapore: Oxford University Press, 1991), pp. 163–64.
25. Wang Gungwu, *China and the Chinese Overseas* (Singapore: Times Academic Press, 1991), p. 171.
26. *Straits Times*, 3 March 1866.
27. Ibid.
28. Lee, *Chinese Society in Singapore*, p. 26.
29. Seah, "The Chinese in Singapore", pp. 283–89.
30. Quoted in Donald Moore and Joanna Moore, *150 Years of Singapore* (Singapore: Donald Moore Press, 1969), p. 202.
31. *Report on the Administration of the Straits Settlements During the Year 1858–59*, p. 55.
32. Ibid.
33. Ibid., p. 60.
34. James Finch, *Engineering and Western Civilisation* (New York: McGraw Hill, 1951), Chapter 7.
35. R.A. Buchanan, "Institutional Proliferation in the British Engineering Profession, 1847–1914", in *Economic History Review* 38 (1985): 42–60.
36. R.S. Cowan, *A Social History of American Technology* (New York: Oxford University Press, 1997), p. 138.
37. Oliver C. Walton. "Officers or Engineers? The integration and status of engineers in the Royal Navy 1847–60", in *Institute of Historical Research* 77 (2004): 178. Walton examines the social dimension of the naval engineer serving in the Admiralty, especially how they faced up with the established ethos of naval and military officers. He argues that the naval engineer played a significant career role in the development the Royal Navy institution, as it made the transition to steam propulsion.
38. R.A. Buchanan, "The Diaspora of British Engineering", in *Technology and Culture* 27 (1986): 501–24. Buchanan traces the diaspora and roles of

British engineers in the "systematic colonization" of places like Australia, New Zealand, India and Africa.
39. Ibid., pp. 522–23.
40. http://home.ozconnect.net/tfoen/singaporejurors1.htm.
41. Norman Edwards, *The Singapore House and Residential Life, 1819–1939* (Singapore: Oxford University Press, 1991), p. 31.
42. Ibid., pp. 132 and 216.
43. Two of Thomson's most popular accounts of his experiences in Singapore and Southeast Asia are *Some Glimpses into Life in the Far East* (London, 1865), and *Sequel to Some Glimpses into Life in the Far East* (London, 1865).
44. Edwards, *The Singapore House*, p. 212. In 1824, Joseph Aspdin, a bricklayer and mason in Leeds, England, took out a patent on a hydraulic cement that he called portland cement because its colour resembled the stone quarried on the Isle of Portland off the British coast. Aspdin's method involved the careful proportioning of limestone and clay, pulverizing them, and burning the mixture into clinker, which was then ground into finished cement. Portland cement today, as in Aspdin's day, is a predetermined and carefully proportioned chemical combination of calcium, silicon, iron, and aluminum.
45. C.B. Buckley, *An Anecdotal History of Old Times in Singapore* (Singapore, 1902; reprinted, Kuala Lumpur, 1965), p. 571.
46. The information here is taken from Thomson's own description of the construction of the Horsburgh Lighthouse in the Straits of Singapore. See J.T. Thomson, "Account of the Horsburgh Lighthouse", in *Journal of the Indian Archipelago and Eastern Asia* 6 (1853): 376–98.
47. T. Chance and P. Williams, *Lighthouse: The Race to Illuminate the World* (London: New Holland Publishers, 2008), p. 183.
48. David and Thomas Stevensons of Edinburgh (popularly known as "lighthouse Stevensons") were well-known engineers responsible for the many lighthouses around the Scottish coast. Acting as agents of technology transfer, David and Thomas Stevensons, together with another British engineer, Richard Henry Burton were engaged by the Meiji Government to build the "Japan Lights" in 1867. Similar to the building of the Horsburgh Lighthouse, Burton, who was trained by the Stevensons and did most of the construction, used granite and native wood for the basic structure and imported from the workshops of the Stevensons reflectors, reflector-frames, lanterns, machines and other sophisticated items. See O. Checkland, *Britain's Encounter with Meiji Japan, 1868–1912* (London: Macmillan Press, 1989), pp. 45–48.
49. *Singapore Free Press*, 3 October 1851.
50. *Annual Report on the Administration of the Straits Settlements During the Year 1856–57*, p. 16.

Pioneers of Change

51. Chance and Williams, *Lighthouses*, p. 211.
52. Buckley, *An Anecdotal History*, p. 420.
53. Ibid., p. 452.
54. Adrian G. Marshall, *The Singapore Letters of Benjamin Cook 1854–1855* (Singapore: Landmark Books, 2004), p. 99.
55. R. Braddell, *The Lights of Singapore* (Methuen & Co., 1934; reprinted, Kuala Lumpur: Oxford University Press, 1982), p. 66.
56. *Annual Report of the Administration of the Straits Settlements During the Year 1886*, p. 160.
57. *Annual Report on the Administration of the Straits Settlements During the Year 1876*, p. 181.
58. H.E. MacCallum, Acting Surveyor-General, Singapore. Paper laid before the Legislative Council on 30 December 1880 and 26 September 1884 on the *State of Survey of the Straits Settlements*, Mss. Ind. Ocn. s. 87, Vol. 77–88.
59. *Straits Times*, April 1914, as reprinted in *Straits Times*, 7 April 1994.
60. Edwards, *The Singapore House*, pp. 220–22.
61. Shin Hirose. "Two Classes of British Engineers: An Analysis of Their Education and Training, 1880s–1930s". *Technology and Culture* 51 (2010): 389.
62. *Annual Report on the Administration of the Straits Settlements During the Year 1855–56*, p. 18.
63. Ibid.
64. Walter Makepeace, "The Port of Singapore", in *One Hundred Years of Singapore*, edited by Walter Makepeace, Gilbert E. Brooke and Roland St. J. Braddell (London: John Murray, 1921; reprinted Singapore: Oxford University Press, 1991) Vol. 1, pp. 578–92.
65. The company's advertisement of its facilities appeared in the *Singapore and Straits Directory*, 1904, p. 116.
66. A. Wright and H.A. Cartwright, eds., *Twentieth Century Impressions of British Malaya: Its History, People, Commerce, Industries, and Resources* (London, 1908; reprinted, Singapore: Graham Brash, 1989) p. 232.
67. G.E. Bogaars, "The Tanjong Pagar Dock Company, 1864–1905", *Memoirs of the Raffles Museum* 3 (1956): 149.
68. Quoted in Makepeace et al., eds., *One Hundred Years of Singapore*, Vol. 1, p. 579.
69. Makepeace et al., eds., *One Hundred Years of Singapore*, Vol. 1, p. 315.
70. J.H. Drysdale, "Awakening Old Memories", in Makepeace et al., eds., *One Hundred Years of Singapore*, Vol. 2, p. 539.
71. Ibid.
72. Fraser and Neave Limited, *1883–1983: The Great Years*, n.d., p. 8.
73. Ibid.

74. *Straits Times*, 27 January 1898.
75. Ibid.
76. F. Swettenham, *British Malaya* (London, 1948; reprinted, London, 1955), p. 281.
77. Ibid.
78. Amarjit Kaur, *Bridge and Barrier: Transport and Communications in Colonial Malaya, 1870–1957* (Singapore: Oxford University Press, 1985).
79. Quoted in Allen, *British Empire*, p. 614.
80. Headrick, *Tentacles of Progress*, p. 9.
81. Ibid. On the issue of diffusion of technology and enclave development, see Inkster, *Science and Technology in History*, pp. 57–59.
82. David Cannadine, *Ornamentalism: How the British Saw their Empire* (London: Penguin Books, 2001), p. 123.
83. Ibid., pp. 122–23.
84. Adas, *Machines as the Measure of Men*, p. 15.
85. G.J. O'Grady, *Being a True and Unbiased Account of the Life and Work of a Civil Engineer in Malaya, February 1928–February 1942. Mss Indian. Ocean, R6, 1942*, pp. 256–58.
86. Rupert Emerson, *Malaysia: A Study in Direct and Indirect Rule* (New York: Macmillan, 1937; reprinted Kuala Lumpur: University of Malaya Press, 1964). p. 488.
87. Quoted in J.G. Butcher, *The British in Malaya, 1880–1941: The Social History of a European Community in Colonial Southeast Asia* (Kuala Lumpur: Oxford University Press, 1979), p. 53.
88. Ibid.
89. Robert Heussler, *British Rule in Malaya: The Malayan Civil Service and Its Predecessors 1867–1942* (Oxford: Clio Press, 1981), p. 15.
90. Allen, *British Empire*, p. 482.
91. Ibid., p. 552.
92. The main source here is the *Singapore and Malayan Directory* for the year 1922 and 1940. It provides a wealth of details on the commercial life of the colony.
93. Similarly, for those thousands of educated Chinese immigrants in the United States who had acquired university degrees in American universities, faced difficulty in trying to secure positions at large Caucasian-controlled, engineering, architectural and scientific corporations during the first few decades of the twentieth century. See Iris Chang, *Chinese in American*, p. 185.
94. G.J. O'Grady, *Life and Work of a Civil Engineer in Malaya*, p. 222.
95. In the British-occupied Shanghai, one of the most blatant practices of racial prejudice was the barring of dogs and Chinese from the Huangpu Park, the

public gardens opposite the British Consulate in Shanghai. The late kung-fu star Bruce Lee demonstrated the anger and frustration of the Chinese of this racial animosity in his movie, *The Fist of Fury*.
96. Allen, *British Empire*, p. 539.
97. *British Malaya*, Vol. 6, No. 9, January 1932, p. 249.
98. Ibid.
99. *British Malaya*, Vol. 7, No. 3, July 1932, p. 62.
100. *Singapore and Malayan Directory 1922*.
101. S. Maaruf, *Malay Ideas on Development: From Feudal Lord to Capitalist* (Singapore: Times Books, 1988), p. 43.
102. Buchanan, *Diaspora of British Engineering*, p. 518.
103. G.J. O'Grady, *Life and Work of a Civil Engineer in Malaya*, p. 222.

3
Maritime Technology and Development of the Port

A close relationship existed between the expansion of the British Empire and British scientific exploration of lands beyond. In the nineteenth century, geography became a significant "imperial science" in British schools and universities. Exploration and map compilation in far-flung regions remained a source of enormous public interest but as geography developed in to a professional discipline, it "began to expand beyond its original focus on exploration and topographical map-making to assume intellectual authority over a wide range of regionally specific environmental, economic, social, political and cultural evidence".[1] Both the amateur explorers and the university-based scholars provided the geographical knowledge necessary for overseas conquest and colonization.[2] The Empire, with its extensive material resources and geographical reach, allowed travellers and scientific explorers to sail into nooks and corners of lands in the periphery and pursue their interests on a genuinely global scale. The expeditions of James Cook, Joseph Banks and Joseph Conrad resulted in huge collections of scientific information which contributed to the development of British science. The gathering of such knowledge relied on a set of institutions closely linked to the imperial government, including the Admiralty, the Hydrographer's Office, the East India Company, the Royal Engineers, the Ordinance Survey, the Geological Survey and the Kew Gardens. The island of Singapore appeared on a variety of regional maps with the Europeans making inroads into the seas and lands of Southeast Asia from the sixteenth century.[3] Charting of the Straits of Malacca and the adjacent coastline, including the Straits of Singapore, was an important preoccupation for the Portuguese, the

Dutch, and, the English. "[B]y the 1550s or so map-making had begun to catch up with the pace of exploration and discovery in South-east Asia"[4] and cartographers, hydrographers, naturalists and explorers of many European nationalities, such as Ferdinand Magellan, Francis Drake, Joris van Spilbergen, Jan van Linschoten, William Dampier, James Cook, John Crawfurd and James Horsburgh contributed immensely to the knowledge of the region depicted on contemporary maps. It is not surprising then that Stamford Raffles was armed with the geographical knowledge of the region which allowed him to make critical choices in setting up a British station. Soon after the island was ceded to Britain in 1819, Captain Franklin surveyed the southern coastline. The chart was used by John Crawfurd to indicate the geographical perimeter of British influence when he formally took possession of Singapore on 2 August 1824. Exploration of lands way beyond Europe was made possible by advances in maritime technology.

Since the late eighteenth century, British engineers were in the vanguard of a revolution that harnessed the power of steam and the strength of iron to create the British Empire and to transform the world economy. In the latter half of the nineteenth century, the world witnessed significant technological breakthroughs in maritime technology.[5] Ancient wooden sailing was to be entirely replaced by the modern vessel built of steel and driven by steam. Britain played a leading role in the changes which took place. In the process, the industrial producers and merchants created the modern shipping industry which enabled Britain to "rule the waves" till the 1920s. From her Osborne House, on the Isle of Wight, Queen Victoria was able to look towards Britain's naval base at Portsmouth, then the largest in the world and an imposing manifestation of the country's sea power. In 1860, the Queen would have been able to detect the H.M.S. *Warrior* — steam-driven, "iron-clad" in five inches of armour plate and fitted with the latest breech-loading, shell-firing guns. The *Warrior* was the world's most powerful battleship, and built as a response to the launched of the French warship *La Gloire* in 1858. Britain had about 240 ships, crewed by 40,000 sailors who proudly represented the world's biggest naval force at this time.[6] And to top it all, Britain owned roughly a third of the world's merchant tonnage. Britain's share of the world's steam tonnage rose from 24.3 per cent in 1840 to about 40 per cent in 1910. Such was the naval supremacy of Britain and Singapore was destined to play a key role in supporting it in the East.

ADVANCES IN MARITIME TECHNOLOGY

It was obvious to British shipbuilders that if the steamship were to compete effectively with sail, its cost of production would have to be reduced and its efficiency improved appreciably. Technical progress in two main areas, namely, the method of propulsion and materials for shipbuilding, eventually won the day for the steamship. The perfection of the compound engine in the 1860s marked the first major breakthrough in marine engineering. The double-cylinder engine allowed higher pressures to be generated and produced greater power from a given unit of steam.[7] At the same time, wooden paddle wheels were replaced by iron screw propeller. But steamship still consumed large quantities of coal and its carrying capacity was about 16 per cent than that of a sailing ship of the same size.[8] In the late 1870s, another engine was added to produce the triple-expansion engine which was considered as the final blow to the dominance of sailing vessels.[9] The parallel development in the material used for steamship building was seen in the shift from wood, first to iron and then to steel. In the 1850s, however, iron steamship was still rare. Though many difficulties relating to iron construction were being solved by the 1860s, the age of the "ironclad" was short. The British navy was already constructing steel ships in the 1870s. Steel saved one-fifth in the thickness of the metal and a 15 per cent saving in weight.[10] Imperfection in steel manufacturing using the Bessemer process was overcome by the introduction of the Siemens-Martin process refined in the 1870s. Therefore, before the end of the century, steel and the triple-expansion engine produced a shipbuilding boom as shipping lines rushed to replace their now obsolete iron-hulled, compound engine steamers.

The rise of the British shipbuilding industry during the period 1870 to 1910 provided a considerable stimulus to economic development in general and, in particular, in the development of newer regions within the British Empire. The immediate effect was the construction of new port facilities.[11] Steamers were built bigger than ever before and virtually no pre-1850 dock in Britain — and elsewhere in the world — could handle them. The most spectacular changes in size and speed of steamships took place in passenger liners. Within a short span of thirty years or so (between the 1870s and 1900s), trans-atlantic steamers more than trebled in length and cut travelling time for crossing the Atlantic from about nine to only four days. Freight charges also fell drastically and this enabled the movement of large volumes of bulky commodities to take place between countries and especially between Britain and her colonies at the "periphery".

Although by the 1860s the steamship was beginning to make serious inroads into the traffic of sailing ships, it did not really come into its own until the opening of the Suez Canal on 17 November 1869. The opening of the water gateway heralded a new era in port development and maritime engineering. Cut through the Egyptian desert and linking the Mediterranean to the Gulf of Suez, the 160-kilometre canal separated Africa in the west and Asia in the east. Its impact was far-reaching. The Canal was designed for steamships, which could sail through it without being towed, while sailing ships continued with the old route via Cape of Good Hope. It provided impetus for the construction of new lighthouses in India, Southeast Asia and Japan. Far from damaging Britain's interests, the Canal gave British shipping a huge boost. Of the 142 vessels that passed through the waterway in February 1878, 112 were under the British flag. Sailing time from England to Calcutta was reduced from forty-five days to thirty days. A Bombay newspaper reported on 21 April 1885:

> There is no doubt the trade of the East with Europe has been revolutionised during the past ten to twenty years and the opening of the canal has been the most important factor in this. Let us not therefore be too parochial minded. If Great Britain has had to suffer some partial and perhaps temporary disadvantage from the canal, Greater Britain, India, Ceylon, the Straits [Singapore], and Australia has a permanent, a great, an increasing gain, not only from an economic but also from a political point of view".[12]

This international waterway was described by the *Straits Times* as "one of the greatest undertakings of modern times ... and, for the first time, the waters of the West shall combine with the waters of the East to bear the burthen of many a keel across the one time desert of the Isthmus".[13] It was indeed a technological marvel that gave a "big push" to world trade.[14] The great increase in the size of steamers coming out from the shipbuilders in Britain and in the amount of tonnage passing through the Suez Canal and entering seaports in the Indian Ocean and Southeast Asia necessitated the provision of dock and warehouse facilities and loading and discharging equipment on a scale hitherto unknown. Old harbours and ports had to be redeveloped with deepened docks if they were to remain viable. Parallel developments of road and rail transport around harbours also had to take place.[15] In the long run, the demand for ships, docks, warehouses, offices, and other supportive services, created increasing employment and investment opportunities for indigenous and foreign companies. How did the port of Singapore respond to this "shipping revolution"?

It is not an exaggeration to say that without steamship, there is no modern Singapore. As economic historian Gregg Huff puts it, "Great ports like Singapore are seldom, if ever, accidents of history".[16] The status of a coaling station was the most important category of ports-of-call in British experience in the nineteenth century, a category which was new and which required considerable developments in terms of business and finance institutions and port installations. As steamship replaced sail on the ocean routes of the world, particularly after the opening of the Suez Canal in 1869, the demand for coal and the need for coaling stations increased. Singapore's port responded to this need with its rapid transformation of bunkering, victualling, watering and docking facilities, staffed with British technical and engineering and port management expertise. Its expansion, in turn, led to further economic opportunities for the thousands of immigrants and to developments in commercial institutions and supporting services. It gave birth to the rise of the shipping freight services and, to a smaller extent, shipbuilding companies owned by Chinese and European entrepreneurs. The Straits Steamship became a household name as its ships dominated the routes along the west coast of the Malay Peninsula and transported all the tin ore belonging to the Straits Trading Company smelted in Singapore.

FROM A RIVER PORT TO AN IMPERIAL PORT

The economic lifeline of Singapore in the early decades of nineteenth century was the pan-shaped entrance of the Singapore River. Up to the 1840s shipping congregated there and along the bund known as Boat Quay. Recollecting his first sight of the Singapore river port, John Turnbull Thomson, the governor surveyor in Singapore during the years 1841 to 1853, gave an illuminating description of the place:

> We hail the "Queen of the East…[and] soon ran up to the shipping and anchored in British waters. In the foreground busy canoes, sampans, and tongkangs bore their noisy and laughing native crews about the harbour….The roadstead was covered with European, Chinese, and

Malayan vessels, in which constant hum of commerce rang, and the question naturally arises, how was the conglomeration of diverse tongues, creeds and nations held together?[17]

The facilities provided at Singapore's first port were not elaborate. Sailing ships had to be anchored in the roads and as near to the river's mouth as possible. Lighters and sampans, usually built and manned by Indians from the Malabar Coast in India, then transported their merchandise to the range of godowns lining Boat Quay. These warehouses were among the first buildings to bring about a significant change in the landscape of the river port. Loading and unloading of goods was done by coolies whose means of access to the boats was just a long wooden plank. Mechanical or labour-saving devices hardly existed at this time. Many early descriptions by Western travellers referred to the activities at the quay as evidence of the growth of trade. By the 1840s, trade had increased from $11.6 million in 1824 to $20 million in 1840.[18] The once swampy island had emerged as one of the main British ports-of-call along of chain of others, like Aden, Port Said, Karachi, Dakar, Colombo and Hong Kong. To cope with the greatly increased demand arising from the expansion of both coastal shipping and international trade, the size and range of port facilities had to be upgraded. The river port, however, had intractable problems, such as limited space, shallow water, and chronic silting especially after the monsoon season. As early as 1822, attempts were made to clear the waterway by using Indian convicts as a cheap form of residual labour. In 1855, imported dredging machinery was used to halt the problem of silting and degradation of the river. Unfortunately, as reported in the *Singapore Free Press*, the British engineer who was to assemble the dredge died after arrival in Singapore. Henceforth, for the rest of the nineteenth century, only ad hoc efforts were made to clean up the river. While Western technology, in the form of the dredging machine, was used on many occasions during this period — and even built locally by the engineering firm, Tivendale & Company — the results were disastrous because of a combination of factors, such as the lack of government funding, shortage of teak wood to construct the dredging framework, the lack of expertise in dredging operations and, strangely enough, importation of faulty machinery from England.[19] In the meantime, congestion at the river became worst and made navigation along the river a tricky operation. There was an urgent need to look for another waterfront site that would complement the shipping and trading activities of the river port. The answer was found in the New Harbour (renamed as Keppel Harbour in 1900) at Telok Blangah, a largely unexplored site west of the Singapore River. This coastal site offered deep water berthing and better servicing facilities for larger vessels. Francis Hawks, an American travelling to this part of the world in the USS *Mississippi* in the early 1850s, recorded that "the wealthy and enterprising Orient Steam Navigation Company have erected at New Harbour, about two and a half miles from the town, a magnificent depot, comprising wharves, coal-sheds, storehouses, workshops, and other buildings, such as would do credit to any English colonial establishment; and this is no slight praise".[20]

To meet the worldwide expansion of shipping activities, construction of wharf facilities was mainly carried out by the Tanjong Pagar Dock Company, registered on 29 September 1864 with a capital of $300,000. But the wharf construction was not without its technical difficulties. The company engaged the services of George Lyon, a local shipwright, who started embankment and excavation works in May 1864. Horse-drawn carts, which tilted dangerously while moving, were used to transport the earth. The aim was to dam up the seaward ends of the embankments in order to provide dry working conditions all day round. Unfortunately, the retaining wall collapsed soon enough.[21] The directors of the company then decided to employ a fully qualified engineer from England. In September 1865, William J. Du Port arrived from Liverpool — very appropriately because the construction of Liverpool's docks was accomplished by a number of enterprising and inventive engineers, led by the famous of all dock-builders, Jesse Hartley, the creator of Liverpool's Albert Dock complex.[22] Formerly a resident engineer at the Birkenhead Docks, he remained in Singapore for less than three years, during which he directed the construction of the then largest granite graving-dock in the East, the Victoria Dock. A public holiday was declared for the ceremony, and the Victoria Dock was opened for public commercial use on 17 October 1868. The *Illustrated London News* reported on the event:

> Its dimensions will admit of the largest steamers being received; and adjoining it are workshops, under the charge of skilled European engineers, fitted with all the machinery necessary for repairs and the construction of new shafts and other purposes ... The Governor [Sir Harry Ord] and his suite entered the dock in two of the Government steamers, and were received on landing by the directors of the company.[23]

Singapore's reputation as a major port in the British Empire was enhanced. More significantly, however, the Victoria Dock was looked upon as a symbol of Britain's unmatched supremacy of the shipping world. Singapore's natural harbour made that possible. The dock, measuring 450 feet long, 65 feet wide at the entrance and 20 feet deep on the sill, was faced with granite throughout. The working depth of the dock was to cater for most ocean-going steamships plying along the Suez Canal.[24] Ten years after the opening of the Victoria Dock, the Tanjong Pagar Dock Company built its second dry dock, the 496 feet long, 59 feet width of entrance, and 21 feet deep, Albert Dock. Its construction was made necessary by the great increase in volume of steam shipping which called at the port after the opening of the Suez Canal in 1869.

Despite changing shipbuilding technology after the 1850s, steamships still required large amount of coal to operate. The coal problem was particularly acute on long distance voyages such as those to the Far East owing to the lack of coaling stations on route. Britain's naval supremacy depended to a large extent on her various coaling stations strategically located along the world's main sea routes. These stations sprang up as the nineteenth century equivalent of the oil depots today. Exploring the profound role coal has played in human history, Barbara Freese puts it: "There is, though, at least one truth that was more widely understood in the past than it is today — the critical importance of coal in shaping the fate of nations, and of the world as a whole. Coal transport lured the British to the sea, promoting the nation's growth from a small rural nation into a world-class commercial power. The Royal Navy was kept strong largely to protect the coal convoys; and in war time, it seized the coal ships and crews to fight its battles, helping Britain rule the seas".[25] Once again, Singapore fitted in the general pattern of British imperial expansion and control of lands and trade. The colony was quickly transformed into a major coaling station to serve the needs of the British merchants.[26] Coal from Bengal was being used in British steamers in the 1830s and it was also exported to Southeast Asia. The Labuan mines were also exploited in the 1840s but the coal proved to be of very poor quality.[27] And Singapore had no coal of its own. Her coal was "dug out of the earth in the old Wales [and] of the mines of New South Wales — thus completing, as it were, a girdle significant of the extent of the British Industry".[28] Indeed, for the people living in Singapore in 1857, they had the opportunity to witness the arrival of the 229 foot long rigged side-wheel steamer with

12 guns, the USS *Mississippi* which had 230 tons of coal loaded onto her so that she could proceed with her journey to Hong Kong and beyond.[29] On board the ship was none other than Commodore Matthew C. Perry — popularly regarded as the father of the United States' Navy because of his strong commitment to introducing steamships in to the Navy, and as the American who led his "Black Ships" in the opening of feudal Japan in July 1853.

To maintain the efficiency in coaling services, the companies at New Harbour built coal sheds to cater for the storage of coal. The Peninsula and Orient Company had a wharf frontage of about 1,200 feet and coal shed built of bricks and tiled roofs, capable of storing about 20,000 tons of coal or its own use.[30] The Tanjong Pagar Dock Company, on the other hand, had various scattered lots of coal belonging to a large number of individual coal and shipping companies in Singapore. Wooden shed with attap or palm-leaf roofs were erected in 1869 but the great fire of 1877 destroyed them. Twenty new brick sheds with tiled roofs, reinforced with corrugated zinc or iron, were then built to provide storage capacity of up to 45,000 tons.[31] In the eyes of local colonial officials, the port was the best-equipped bunkering and coaling station in the region. In 1884 the Governor of the Straits Settlements, Frederick Weld, heaped praise on the coaling efficiency of the port:

> The facilities of coaling with despatch at New Harbour are almost unequalled; fully 300,000 tons are usually stored there, and labour is plentiful. The famous steamer *Stirling Castle*, racing homewards with tea from China, had 1,600 tons of coal put on board her in four hours, and her rival the *Glenogle*, the same day, 1,800 tons at the same rate. There are facilities for docking all but the very largest ships, numberless ships of war coal and dock there, and I hope to bring about an arrangement between the Admiralty and the Tanjong Pagar Dock Company, by which Singapore will have the largest dock in Asia, capable of docking any of Her Majesty's ships of war.[32]

Weld's comments were indicative of the high labour productivity of "coaling" or loading coal. The coaling rate, that is, the amount of coal measured in tons loaded onto a steamer in an hour, worked out to be about 400 tons per hour. Weld's coaling figures could have been inflated. The documented record for coaling feat at the port took place on 4 August 1902 at Keppel Harbour when 1,510 tons was put on board H.M.S. *Terrible* in five hours or about 300 tons per hours.[33] The *Straits Times* described the world record feat:

> Excitement centred around the H.M.S. *Terrible* yesterday morning when it became known that Messrs. Paterson Simons & Co., the contractors to the British Government, and the Tanjong Pagar Dock Co., were going to attempt a record feat in the coaling of the giant cruiser... The *Terrible* was moored alongside the wharf at 9 a.m. and at 7 minutes past 10 work commenced. The heat was intense and the atmosphere simply full of Cardiff coal dust, but work did not slacken until about 3 p.m. when the high pressure was relaxed for the very simple reason that there were only remained some 48 tons to be got abroad before the vessel's departure. Altogether the 1,548 tons were got in 5 hours and 27 minutes or at the rate of some 302 tons an hour, a really remarkable feat.... It may be noted that no less than eight vessels were working coal at the same time, no less than 5,135 tons being handled during the day.[34]

The port was gaining a reputation of operational efficiency. Its coaling rate was much higher than those at some of Britain's leading ports — 110 tons an hour at Colombo, 90 tons an hour at Aden and 200 tons an hour at Port Said during the period 1890 to 1913.[35] While the port increasingly gained in commercial importance and, in Weld's words, "was acknowledged by the highest authority to be quite one of the most important economically or strategically in the Empire", it was "at present said to be virtually defenceless".[36]

The intensive use of Chinese coolies in the British Empire has been well documented in many contemporary travellers' accounts. One description of the tea plantation in China was provided by the British lady traveller, Constance Gordon Cummings in the 1870s: "I am greatly struck by the number of girls whom we meet working as tea-coolies, and by the enormous burdens which they carry slung from a bamboo which rests on their shoulder. Each girl carries two bags thus slung, the weight of a bag being half a *picul*, which is upwards of 60 lb. Thus heavily burdened, a party of these bright, pleasant-looking young women march a dozen miles or more, chatting and singing as they go..."[37] Interestingly, Western travellers' accounts also attributed the unflagging energy of Chinese coolies (and even accounting for the rapid population increase of China's and Japan's population in the nineteenth century) to the tea-drinking habit.[38] In Singapore, the high labour productivity at the port was due, in no small measure, to the Chinese dock coolies. Coaling itself was a messy and laborious task. It was usually done at night but it was the ship's schedule that was the deciding factor. *The Graphic*, dated 4 November 1876, gave an interesting account of the nature of the job:

> This is entirely carried on by the Chinese population at the port, and is most efficiently performed. A Chinese contractor is seated on a barrel by the line of coolies, having beside him a tray of small coins and a weighing machine. The first man of each group as he approaches the contractor receives from him one cent and a quarter; at which rate 160 basketfuls have to be carried on board before each coolie earns a dollar. About one in every dozen loads is weighted to ensure the proper amount of coal being carried in each basket. The rapidity with which a ship is coaled is wonderful; the coolies work with unflagging energy without cessation all day, in spite of the blazing sun, against which the sole protection they have is a straw hat.[39]

In this fashion, the Chinese coolies performed an invaluable service to the steamers of The Peninsula and Orient Company, the French Messageries Maritimes, the China Company of Liverpool, the Netherlands India, the Austrian Lloyd's and other major shipping lines. The total number of Chinese coolies at the port was not known. But in 1879, the Tanjong Pagar Dock Company alone employed about 2,500 men, most of whom were certainly dock coolies.[40] The figure climbed to 3,000 to 4,000 workers in the 1880s as a result of the increase in number of shipping vessels calling at the port.[41] The directors of the company insisted on retaining the usage of coolie labour and rejected proposals made to introduce labour-saving mechanical appliances.[42] In their views, since the coal sheds were scattered about the harbour area, it was more economical to employ human labour than to purchase and maintain a number of these mechanical devices. The former had greater mobility than fixed machinery. Until 1874, cargo at the Tanjong Pagar wharves was loaded and unloaded manually at the rate of 200–300 tons per day of 10 hours. In that year, steam winches and cranes were installed and working of cargo was increased to 500–800 tons per day. In 1897 electricity was introduced into the entire wharf frontage, roadways and docks, and working hours were virtually doubled. By comparison, ships at the anchorages could only work their cargoes at the rate of 7.5–10 tons per daylight hour.

Besides being supported by a plentiful supply of coal and a huge labour force, the port of Singapore also invested in workshop facilities in order to meet with the more sophisticated steamships that called at the harbour. The construction of large marine engines, much bigger than those required for mills or railroad locomotives, for example, required new and bigger machine tools, in particular, bigger lathes and boring machines. The Tanjong Pagar Dock had a machine shop which was described as "one of the most complete in the East, fitted with lathes, shearing, punching, shaping, slottings, planing, screwing, and boring machines, with the latest improvements, with all the necessary tools and appliances required for effecting the most extensive repairs of steamers and iron vessels of the largest class..."[43] Electric light was first installed in the workshops in 1878.

By the end of the century, with the exception of the Peninsula and Orient Company which had its own wharf, the Tanjong Pagar Dock Company had monopolized the port activities by taking control of all its rivals. However, the maintenance and further development of modern port facilities required massive capital expenditure. The funds could not be easily obtained from the existing port revenues or even from the colonial authority. Besides the financial input, the sheer physical change to the morphology of the port also necessitated the adoption of new forms of administration and management. In Britain, such dynamic changes had led to the formation of the Mersey Docks & Harbour Board in 1857 which revamped and administered the entire harbour infrastructure of Liverpool and Birkenhead.[44] For the Singapore port, the change to autonomous public ownership and administration came in 1905. A year before, the Tanjong Pagar Dock Company had proposed an extensive improvement scheme, involving an estimated expenditure of $12 million. This was felt to be essential since the world was moving into the technological age — steam trains, electric trams, submarine telegraph, and motor cars were creating new and exciting environments in the West. Unfortunately, there was disagreement between the local management and its London Consulting Committee, which was more concerned with short-run maximized profits. Despite being the seventh largest port in the world, measures to inject fresh capital into the modernization of the facilities were never supported by the major shareholders. Even Sir Frank Swettenham, as the Governor of the Straits Settlements, was not able to sway opinion. It was left to his successor, Sir John Anderson who persistently pushed for a policy of expropriation. This forced the British government to intervene. A contemporary observer noted in his memoir:

> One of the first moves made by Sir John Anderson after his appointment as governor was to ask the Legislative Council to assent to the expenditure of a large sum of money on the Singapore harbour.... That it was a good move was the general opinion because it was recognised on all sides that the company was rather out of date. A very large sum would be required to carry out the rehabilitation and modernising of the property which was very necessary and, in fact, overdue.[45]

By 1 July 1905 the property of the Tanjong Pagar Dock Company passed into public ownership (under the Tanjong Pagar Dock Board) and subsequently passed over to the Singapore Harbour Board which was constituted under the Straits Settlement Port Ordinance of 1912. The Singapore Harbour Board started with a staff strength of 2,050 of whom 100 were Europeans.

Large-scale development of the port took place, starting with the construction of the new Lagoon Dock and the reconstruction of the wharves at Tanjong Pagar. The port authorities invited established British engineering firms to bid for the project in 1907 and eventually it was awarded to John Aird & Co. The company, under Sir John Aird, had excellent track records. It was commissioned by the Egyptian Government in 1898 to construct the dams on the Nile at Aswan and Assiout. The project, regarded as one of the most ambitious engineering works carried out in the Middle East, was completed way before schedule and transformed Egypt's ancient irrigation basin to a perennial system. The expectations of the Tanjong Pagar Dock Board, however, were dampened when it was reported in 1910 that "the progress of the Lagoon Dock and Main Wharf Reconstruction, Tanjong Pagar, by Messrs. John Aird & Co., has not been all that had been anticipated with such experienced contractors".[46] A year later, the contractor stopped work altogether, the action of which "will delay the completion of this section of the Dock improvements for some three years, a result which in view of the urgent need of extended wharfage cannot fail to be very prejudice to the port of Singapore and so to the Colony".[47] On 8 March 1911, it was decided that Topham, Jones & Railton, the contractors for the graving docks at Keppel Harbour, would take over the project and the company subsequently commenced work on 18 July 1911. After three arduous years, Singapore's Empire Dock was successfully completed on 2 June 1914 and the first ship to enter was the *Valdura*. In 1920, an electrical crane for the handling of heavy weights at the wharves was added to the range of wharf facilities.[48] Although the Singapore economy suffered during the period of the Great Depression, the Singapore Harbour Board continued to make steady progress with technical changes to the port's infrastructure during the 1930s. By this time, the Board employed about 100 European officers, 1,900 locals as office staff, 4,000 wharf coolies and more than 5,000 skilled Chinese artisans of trades with their assistants and labourers working in the dockyard department.[49] Its warehouses were now built with fire-proof ferro-concrete. All the wharves were expanded and constructed with steel and corrugated iron, concrete floors and with projecting eaves for protection against rain in discharging or loading cargoes. The port was connected directly to Malaya's main railway line by the Board's railway system of some 22 kilometres of metre gauge track. In October 1935, it was decided that a new mechanical handling plant would be added to the existing facilities and the contract was awarded to the British firm, Fraser & Chalmers Engineering Works. After the War in 1945, the high cost of labour resulted in further deployment of modern machinery by the Singapore Harbour Board. A variety of labour-saving equipment was introduced, including the Ransome electric elevating-platform trucks, fork-lift trucks, Nimrod mobile cranes, and Chase side-towing motors.

NEW SHIPPING AND COMMERCIAL ORGANIZATIONS

Although technological changes opened up a new chapter in world shipping and tropical harbours like Singapore and Colombo responded with modernization of their port facilities, technology by itself was not the only catalyst for change. Perhaps even more remarkable than the physical development of ports and of mercantile fleet in the latter half of the nineteenth century were the changes which took place in the structure and organization of the shipping industry. This transformation was also brought about by the expansion of world trade.

There was a gradual trend away from casual ownership and operation of vessels by merchants and individuals towards larger business undertakings which specialized in shipping exclusively. British shipping companies were quick to create new operational changes and regular shipping services in several areas, such as in finance, scheduling procedures, passenger traffic, mail and cargo contracts. As the first steamship line in the East, the Peninsular and Oriental (P&O) Steam Navigation Company monopolised mail services to South Asia and Southeast Asia after 1845.[50] Originally the Peninsular Steam Navigation Company, serving Portugal and Spain, the "Oriental" was added in 1840 when the first regular mail service to India was inaugurated. In 1845, the Company contracted to carry the monthly mail to China via Ceylon, Singapore and Hong Kong. It used two ships at the start — the *Lady Mary Wood* and *Braganza*. On 4 August 1845, the *Lady Mary Wood* became the first steamer to arrive in Singapore, en route for Hong Kong.[51] The *Straits Times* carried a short report: "The Orient and Peninsular Company's Steamer *Lady Mary Wood*, Captain Cooper, arrived at Singapore yesterday morning with the Overland Mails of the 7th and 24th June ... The Steamer brought the following passengers: For Singapore — Dr Mc Questen; For Hongkong — Messrs Skinner and Mills; For Manila — Messrs Geronelle, Torres and Malvar" and it went on to describe some particulars of the mail from London which "are not of much interest".[52] Nevertheless, it was a momentous occasion as letters from friends and loved ones arrived into Singapore in just forty-one days. It also marked the first homeward mail being dispatched to London — or so everybody thought. As the *Lady Mary Wood* sailed out to sea, it was discovered that all the prepaid letters were left behind due to the carelessness of the officials. Despite this blunder and as the anger subsided, it was acknowledged with contentment by the community that "the mail system opened up rapid and regular intercourse with Europe and China and the intermediate ports".[53]

In 1852, P&O was granted a contract for mail service from Singapore to Sydney using the ships, *Chusan* and *Formosa*. Besides mail and passengers, the P&O liners (with seventy-one ships in 1912) also carried precious cargoes, such as frozen English beef to Singapore. The dominance of the P&O Company was soon challenged by the formation of several new liners. The British India Steam Navigation Company was founded in 1856 by William Mackinnon and it quickly prospered through its short distance shipping trade in rice and timber and mail contracts between Calcutta and Southeast Asia. However, the Company's strength laid in its specialization in bulky freight and transportation of migrants between India, Burma, Ceylon, Malaya and China. In 1907, the Company won a contract to transport Indian immigrants to Southeast Asia, and it continued to enjoy this monopoly till the outbreak of war in 1941.[54] In 1865, Alfred and Philip Holt of Liverpool formed the Ocean Steamship Company (the Blue Funnel Line) and, through the use of the most technically advanced ships, gained a large share of the trades of Malaya and Dutch Indonesia. By the end of the nineteenth century, several other major British and Continental shipping lines, such as the Ben Line and the German Norddeutshcer Llyod Company, also became household names in Singapore.

Besides the technical improvements made on the infrastructure of the port at the New Harbour, the presence of these international shipping lines also stimulated the development of new business organizations in the commercial heartland of Singapore — in Collyer Quay, Finlayson Green and Telok Ayer Basin. As happened to coastal shipping companies in Britain in the early nineteenth century, the transition from sail to steam in shipping resulted in the incorporation of joint-stock shipping agencies, some of which are now celebrated names in Singapore's maritime history.[55] One such organization was the shipping agency. The Ocean Steamship Company, for example, awarded its Singapore agency to Mansfield & Company in 1868 and the latter eventually became one of the most successful local shipping agencies.[56] These shipping agencies ensured the expeditious conduct of shipping businesses, such as making arrangements in advance for the loading and shipment of cargoes, the arrival and departures of passengers and meeting the requirements of the authorities for ships docking at the wharves. Joint enterprises between Europeans and Chinese businessmen also took place. Besides the Tanjong Pagar Dock Company, another such enterprise was the Straits Steamship Navigation Company, registered in Singapore on 20 January 1890. It was a collaborative venture between European entrepreneurs and, what Eric Jennings called "The Malacca Connections" of three prominent Straits-born Chinese — Tan Jiak Kim, Tan Keong Saik, and Lee Cheng Yam.[57] The enterprise was the brain-child of a Dutchman, Theodore Cornelius Bogaardt who persuaded Phillip Holt of Blue Funnel to participate in Southeast Asia regional trade.[58] The company was registered in Singapore beginning with a nominal capital of $10,000,000 in $10 shares, its issued capital however was no more than $421,000 and all the shareholders were local participants. According to the *Straits Times*, it began with five ships, *Will O' the Wisp* (148 tons), *Sappo* (324 tons) both contributed by Bogaardt, the remaining three *Hye Leong* (406 tons), *Malacca* (404 tons) and *Billiton* (335 tons) contributed by the Tan family. They were built in Britain in the 1880s and were "schooner rigged with low powered engines and all had British masters".[59] In no time, the company controlled much of the Singapore and Malaya coastal trade.[60] By the 1920s, the company ran more than fifty of these shallow-draught steamers, known as "the little white fleet" on account of their white hulls and blue and white funnels — that ran with clockwise precision between nearly eighty ports in Southeast Asia.[61] Timing, profitability, local conditions, and the quality of leadership appeared to have been important in determining the nature of the response in business strategy for the company. The opportunities to learn and apply Western shipbuilding technology were well tapped and for many years to come, the Straits Steamship Company dominated the Malayan coastal trade.

The growing scale and complexity of trade and shipping needs also inevitably led to a greater specialization of the tertiary sector. Important supporting facilities, such as banking, insurance, shipping agents, and brokers were developed to service the trade-based economy of Singapore. Banking businesses in the late nineteenth century, for example, was still monopolized by the three British banks, namely, the Chartered Bank of India, Australia and China, the Hong Kong Shanghai Banking Corporation, and the Mercantile Bank of India. However, at the turn of the new century, the financial sector saw the opening of several new banks. The Nederlansch Indische Handelsbank, a Dutch bank, opened in June 1901; the International Banking Corporation, an American bank, came to Singapore in 1903; the Kwong Yik Bank was set up by local businessmen in 1903; and the Singapore branch of the French bank, Banque de Indo-Chine was established in 1904. The number of foreign firms hogging the commercial buildings in the city also increased steadily, from fourteen in 1827, to thirty-six in 1855 and increasing to sixty-two in 1872. Although principally involved in the shipping and entrepôt trade, by the beginning of the twentieth century, many of Singapore's established agency houses — such as the Borneo Company Limited, Guthrie & Company Limited, Henry Waugh & Company Limited, Edward Boustead & Company, and Harper, Gilfillan & Company Limited — diversified their interests into managing and owning estate plantations, tin mining industries and other light industries. The European agency houses had become widely known in the British Empire and possessed considerable capital. Supported by their involvement in various enterprises, the agency houses could maintain knowledgeable staff, including local English-educated staff. Efficiency and continuity of operations were thus secured. The multiplier effects created by the arrival of major liners into the port of Singapore were indeed wide ranging.

THE ROLE OF THE PORT IN THE BRITISH EMPIRE

During the last quarter of the nineteenth century, more than half of the world's merchant shipping flew the Red Ensign.[62] The British merchant imperialists had a tenacious grip on world trade because supplies of foodstuffs and raw materials from all over the globe were vital to an industrialized Britain. To protect the imperial seaways, elaborate systems of supply, Britain's key ports and maritime fortresses all over the world operated an inter-connected communication and defence system. In Southeast Asia, Singapore was the beacon of British economic and political power. The island was regarded as the centre of "the Golden Chersonese", described by Frederick Weld, as "a circle drawn round Singapore with a radius of 3,000 miles, [which] is believed to contain more than half the population of the globe, and Her Majesty's possessions within this range are stated to have a sea-going trade of 251,000,000 pound sterling, against 86,000,000 pound sterling in all other British dependencies".[63] The port was Britain's key port of calls for stores and repairs of its imperial naval and commercial fleet. More significantly, Singapore sustained the Empire's trade along the main oceanic routes by operating as its major coaling station in the East. Here, it was possible for steamers to be supplied with Australian, Indian, Japanese or Welsh coal.

The growth and rapid development of modern, imperial ports such as Singapore (and Madras, Bombay, Calcutta and Colombo), not only facilitated Britain's world trade and commodity flows within the Empire, they also contributed to the strengthening of imperial control over the exploitation and export of raw materials, minerals and foodstuffs from the colonial economies. Inevitably, the question of military protection of these important ports was a constant issue debated by parliamentarians in London. In the 1890s there was growing fear that a situation could develop in which the navy simply could not manage the task of protecting the British merchant marine.[64] This uneasiness was also partly due to widespread influence of the writings of the American navalist A.T. Mahan who stressed the important part played by a battle fleet in maintaining commercial and imperial supremacy.[65] Imperial Germany under Admiral Tirpitz had been accelerating its naval expansion while the United States and Japan had started constructing powerful fleets for the first time. By the close of the nineteenth century, the might of the British navy was seriously challenged by other imperial powers in different parts of the trading world. The issue of Singapore being developed as a naval base capable of docking any of Her Majesty's ships of war was made more urgent by the geopolitical situation in Southeast Asia after the 1870s. The British were faced not only with the ambitious intentions of new would-be colonial powers, but renewed military activity by the French in Indo-China. In the 1880s and 1890s the French embarked on a series of campaigns to consolidate their control and influence over Vietnam, Laos and Cambodia. In the East Indies the Dutch extended their control over the outer islands; the Spaniards strengthened their position in Mindanao, only to be displaced soon enough by the Americans. Desperately hoping to preserve her independence, Siam also became a theatre of political negotiations between the various colonial powers. Further away in the Far East, Japan was making its impact in international politics. In the face of these major changes, Britain had to adjust and adapt in order to preserve her economic interests in this part of the world.

Singapore's importance to the strategy and trade of the British Empire in the East was reiterated by J.C.R. Colomb, the younger of the Colomb brothers, founders of the "blue-water school" in a conference of prominent colonial officials held in London in June 1884:

> You are also aware that recent events have placed at the other side of Singapore a powerful fleet of transports and a tolerably large French force. If in the outcome of a situation which looks gloomy enough you were suddenly involved in a war, I ask you to consider what would become of the vast store of coal at Singapore, of the Queen's representative, of the Government, and of all the civilisation of which we have been the pioneers in those parts. I say, without fear of contradiction, that if you continue to neglect and to leave defenceless these keys of Empire, you must expect to lose suddenly, in some parts of the world, your supremacy of the sea. I will ask you to consider the world-wide importance of Singapore as a strategic position with regard to commerce, and to turn to a map of the world and consider the connection between Singapore and North America and the value of Singapore as regards Australia and

India, and the guarantee which the safe holding of that place gives for the maritime peace of the world.[66]

The comments made by Colomb reinforced the argument that Singapore's growth, and for that matter, her modern existence, was contingent upon the way British statesmen view the larger economic and geopolitical framework of Southeast Asia and of the empire as a whole, and how Singapore could fit into the grand schema of imperial trade and expansion.

As the nineteenth century came to a close, Singapore did not possess a naval base compatible to her "worldwide importance". This was because the nineteenth century British Liberals and Radicals "saw themselves constructing a cosmopolitan order based on free trade and an international division of labour, and on Christian ethics".[67] The prevailing ethos and ideology belonged to the so-called Manchester School of Radicalism. The "Manchester School", the term first used by Benjamin Disraeli ((British Prime Minister in 1868 and from 1874 to 1880) in February 1846, was in fact a mixed group of businessmen, traders and manufacturers from all parts of Britain who supported free trade because of the prospect of lucrative profit. But immediate gains were not forthcoming because Lord Palmerston, as Foreign Secretary and then as Prime Minister between the years 1830 to 1865, was not an active free trade promoter.[68] To radical politicians like Richard Cobden and John Bright, capital not conquest was the true foundation of national prosperity. Reiterating the Free Trade thinking, the philosopher and jurist Jeremy Bentham commented that "all trade is in its essence advantageous", and "all wars are in its essence ruinous".[69] In the context of this prevailing ideology, there was no need to maintain an all-powerful navy to protect Singapore. The port thus remained as a free trade centre way into the new century. It was only after the First World War, when *laissez-faire* and free trade came under heavy attack in Britain and when the aftermath of the war was fully absorbed that plans were made by the politicians to build the ill-fated naval base in Singapore.

CONSTRUCTING THE ILL-FATED NAVAL BASE

The history of the Singapore Naval Base has its beginning in 1919 when Admiral Jellicoe was sent to the Pacific and Far East to report on the future naval strategy in those regions. Of his various recommendations, the most important was the establishment of a large Far Eastern Fleet and stationed upon a new base at Singapore. A site of about 4 square miles at the extreme north of Singapore, fronting the Straits of Johore, was designated as the naval stronghold of the British Empire in Southeast Asia. It was a scarcely inhabited, forested and malarial-infested land. However, ground preparations were consistently checked by the change of governments in Britain, with their respective policies relating to the stationing of a British fleet in the Far East. Economic considerations were also significant.[70] In 1921 the Lloyd George government announced that work would begin on the naval base — but in reality the work did not take off. By the early 1920s, the British economy was showing signs of stress, with a high rate of unemployment and decline in the export industries. There was unprecedented public demand for defence cuts. In 1924, under Ramsay MacDonald's pacifist government, the site was left very much undeveloped. Excavation and piling works were restarted in November 1924 (by a Conservative Government) but was stopped again in 1929 with the Socialist Government in power in the British Parliament, when "work was slowed down as much as possible, and a 'truncated scheme' adopted under which completion of the dockyard and defence works, apart from the current contracts, was to be postponed for five years".[71] The transformation work to the tract of land was enormous and demanding. Mangrove forest enmeshed in dense foliage was cleared. Millions of tons of earth were excavated and 34 miles of concrete and iron piles were driven through mephitic swamp to meet bedrock at a depth of 100 feet. Graving docks, capable of taking the biggest ships, and huge cranes, including a self-propelled floating crane and capable of lifting about a hundred tons, were dug out and shore installations were into place.[72] The Singapore Naval Base was ready to handle any warship likely to be built in the future. *The Times* aptly described the change in 1938:

> In 1922 the site of the dockyard was jungle, malarial swamp, and rubber plantations. This has now been transformed into a healthy modern dockyard, having over 5,000 ft. of deep-water quays, with 30 ft. to 40 ft. alongside at low water, a great 1,000 ft. graving dock [named as King George VI] which can take the largest ship in the world, workshops, power stations, storehouses, hospitals, and living quarters.... A floating dock, also capable of lifting the largest ships, was placed in position in 1928, and there is also a floating crane capable of handling the enormous weights with which a naval dockyard deals. All the civil engineering work,

including the building of the graving dock, has been carried out by the form of Sir John Jackson, Limited.[73]

As a symbol of British naval supremacy in this part of the world, the Singapore Naval Base boasted of fortifications that were "capable of repelling seaborne attack by any force, however strong ... with guns not only of 15 in. but of 18 in."[74] Inside the base, ringed by high walls, iron gates and barbed wire, were barracks, offices, stores, workshops, boiler-rooms, refrigeration plants, canteens, churches, cinemas, a yacht club, an airfield and seventeen football pitches. There were huge furnaces, vast crucibles and troughs for molten metal, enormous hammers, lathes and hydraulic presses. Underground storage tanks were built to contain a million tons of fuel and a new torpedo depot was being built as late as 1941. Arising from the construction of the base, a range of supporting infrastructure and commercial services were also developed in the vicinity. Machinery workshops, storehouses, hospitals and barracks were built and came with the latest in modern sanitations.

The Singapore Naval Base was officially opened by the Governor, Sir Shenton Thomas, on 14 February 1938. It was a proud occasion for Sir John Jackson and his entourage of engineers and contractors. For the thousands of British military personnel and civilians living at the periphery of the British Empire, in an island once dominated largely by fishermen, the Singapore Naval Base reassured them of the might of the British Empire. The Governor proclaimed the completion of the naval base as "a great enterprise for peace begun nearly 15 years ago". At 5.30 p.m. the Governor's yacht, the *Seabelle*, sailed into the gigantic dock, escorted by the warships *Duncan*, *Diamond* and aircraft carrier, H.M.S. *Eagle*. Britain's main ally, the United States, was represented by a naval squadron consisting of the cruisers *Trenton*, *Milwaukee*, and *Memphis*. As reported in *The Times*:

> The base is no challenge to war but an insurance against war. It has grown up slowly, in spite of difficulties, the symbol of the care of the Mother Country for the welfare of the people and their protection in time of need. Other countries have watched the development of the base with sympathy and satisfaction, for the British Empire is one of the most potent influences on civilization that history has ever known. Built up on the foundations of truth, liberty, and justice, united in a common bond of loyalty to the Throne, it needs one thing: It must be strong.[75]

The base was ready in 1941. Britain only needed to provide the so-called Eastern Fleet. Ironically, on the fourth anniversary of the inauguration of the Singapore Naval Base, it was an entirely different script — the base was a scene of self-destruction and mayhem. The British Empire suffered one of its worst humiliations when its imperial port-city of Singapore fell, in words of Sir Alexander Cadogan, the permanent undersecretary of the British Foreign Office, to the "beastly little monkeys" from the Land of the Rising Sun.[76] On the eve of the British debacle at Singapore, the British humour magazine *Punch* depicted Japanese soldiers in full-page splendour as chimpanzees with helmets and guns swinging from tree to tree. Earlier on 15 May 1940, Winston Churchill wrote his first prime-ministerial letter to President Roosevelt. In it, he emphatically stated that: "I am looking to you to keep that Japanese dog quiet in the Pacific, using Singapore in any way convenient".[77] The ill-fated Singapore Naval Base, seen as a symbol of British pride and strength, had fallen.[78] Singapore had fallen.

Notes

1. Morag Bell, R Butlin, and M. Hefferman, eds., *Geography and Imperialism 1820–1940* (Manchester and New York: Manchester University Press, 1995), p. 5.
2. Ibid., p. 6.
3. R.T. Fell, *Early Maps of South-East Asia* (Oxford and New York: Oxford University Press, 1988).
4. Ibid., p. 9.
5. While the Europeans led the world in maritime technological innovations during the eighteen and nineteenth centuries, it should be noted that from the eleventh through the fifteenth centuries, China held sway in shipbuilding. Chinese ships of this period were nearly 200 feet long (as compared to European vessels which were less than 100 feet long), had rudders attached to their sterns (an Eastern innovation that took centuries to reach Europe) and watertight bulkheads reinforced their decked-over hulls, an innovation that even the builders of the *Titanic* did not quite get right. See James Delgado, *Kamikaze: History's Greatest Naval Disaster* (London: Vintage Books, 2009), Chapter 2.
6. Today, China is doing what the Victorian Royal Navy did to protect its imperial sea routes during the last quarter of the nineteenth century. In the words of Rear Admiral Zhang Huachen, Deputy Commander of the East Sea Fleet: "With the expansion of the country's economic interests,

the navy wants to better protect the country's transportation routes and the safety of our major sea-lanes". See Elizabeth C. Economy, "The Game Changer: Coping with China's Foreign Policy Revolution", *Foreign Affairs* 6 (November/December 2010): 149.
7. Headrick, *Tentacles of Progress*, pp. 29–30.
8. H.J. Dyos and D.H. Aldcroft, *British Transport: An Economic Survey from the Seventeenth Century to the Twentieth* (Harmondsworth: Pelican Books, 1969), p. 259.
9. Ibid., pp. 259–60.
10. Headrick, *Tentacles of Progress*, p. 29.
11. See Adrian Jarvis, ed., *Port and Harbour Engineering* (Aldershot: Ashgate, 1998). The contributors in this volume provide a wide coverage on the development of major ports in Britain and the evolving port engineering that had to be constantly improved and adapted especially after 1850 with the introduction of the steam engine and the opening of the Suez Canal in 1869.
12. Quoted in T. Chance and P. Williams, *Lighthouse: The Race to Illuminate the World* (London: New Holland Publishers, 2008), p. 198.
13. *Straits Times*, 17 November 1869.
14. While the opening of the Suez Canal in 1869 marked another milestone in technological progress, just a decade ago, in 1858, the opening of the Egyptian railway created a profound impact on the existing infrastructure of travel and tourism in the Nile Delta and the Western and Sinai deserts. A line connecting Alexandria to Suez, built in 1851–58 with government funding, inaugurated the rapid expansion of railways in Egypt. By 1914 the country had a railway network of about 4,300 kilometres, which was carrying the bulk of the internal goods traffic. Prior to the arrival of the "iron-horse", cargoes (including water and coal) were off-loaded onto the backs of thousands of camels and trundled across the desert to Suez in the South or to the Mediterranean ports in the North. There was also an elaborate system of Nile steamers, coaches and rest houses, all catering to the weary travelers leaving the waters and transported across the deserts to their destined ports. Railway technology put an end to all this.
15. In Britain itself, for example, the port of London, Liverpool and Glasgow built a whole series of specialized docks and equipped them with the latest machinery in order to deal with the wide range of commodities which entered these ports. In 1828 the largest vessel which the port of London had to accommodate was the *British Queen* of 2,016 tons gross but, by 1900, the *Celtic* belonging to the Oceanic Steam Navigation Company was more than ten times the size (20,904 tons).
16. W.G. Huff, *The Economic Growth of Singapore: Trade and Development in*

the Twentieth Century (Cambridge: Cambridge University Press, 1994), p. 3.
17. J.T. Thomson, *Glimpses Into Life in Malayan Lands* (Singapore: Singapore University Press, 1984), p. 14.
18. Wong Lin Keng, "Singapore: Its Growth as an Entrepôt Port, 1819–1914", *Journal of Southeast Asian Studies* 9 (1978): 57.
19. Stephen Dobbs, *The Singapore River: A Social History 1819–2002* (Singapore: Singapore University Press, 2003), pp. 53–54. It was only in 1884, with the arrival of the new bucket dredge, powered by a steam engine, that extensive clearing operations finally took place.
20. L.H. Francis, *The Singapore Chapter of the Narrative of the Expedition of An American Squadron to the China Seas and Japan* (New York: D. Appleton & Company, 1857; reprinted, Singapore: Antiques of the Orient, 1988), p. 5. Linking the Keppel wharves with the warehouses on the Singapore River were two earth roads along which the bulk of the cargo was transported by bullock carts. When an increase in the volume of cargo followed the opening of the Suez Canal, a state of serious congestion resulted at the wharves and along the roads. This led to the first phase of the Telok Ayer reclamation in 1879 which, when completed in 1887, extended the foreshore from Telok Ayer Street to Raffles Quay and provided an additional 18 acres of land on which the new access roads between Keppel Harbour and the Singapore River were constructed. At the same time Keppel Road was laid across the mangrove swamps from Tanjong Pagar (later converted to an electric tramway) also ran on this route.
21. *Straits Times*, 14 May 1873.
22. Sheila Marriner, *The Economic and Social Development of Merseyside* (London: Croom Helm, 1982), p. 32. Besides cast-iron and bricks, Hartley's favourite material was granite blocks which were frequently fitted in random jig-saw patterns to produce massive walls and huge columns.
23. *Illustrated London News*, 2 January 1869, as quoted in D.J.M. Tate, *Straits Affairs: The Malay World and Singapore, Being Glimpses of the Straits Settlements and the Malay Peninsula in the Nineteenth Century as seen through The London Illustrated News and other Contemporary Sources* (Hong Kong: John Nicholson Ltd., 1989), p. 8.
24. G.E. Bogaars, "The Tanjong Pagar Dock Company, 1864–1905", *Memoirs of the Raffles Museum* 3 (1956): 129.
25. Barbara Freese, *Coal: A Human History* (London: Arrow Books, 2003), pp. 251–52. Freese provides a fascinating story of the impact of coal in human history, from its role in shaping industrial technology and the growth of cities like London and Manchester during the era of the Industrial Revolution to its role in China's historical development.

26. W.E. Minchinton, "British Ports of Call in the Nineteenth Century", *The Mariner's Mirror* 62, no. 2 (1976): 151.
27. Headrick, *Tentacles of Progress*, pp. 44 and 282.
28. J.R. Colomb, quoted in Paul Kratoska, *Honourable Intentions: Talks on the British Empire in South-East Asia Delivered at the Royal Colonial Institute, 1874–1928* (Singapore: Oxford University Press, 1983), p. 80.
29. Francis, *Expedition of An American Squadron*, p. 5. The *USS Mississippi* and her sister ship, the *USS Missouri*, were the U.S. Navy's first ocean-going side-wheeled steamers.
30. John Cameron, *Our Tropical Possession in Malayan India* (London: Smith Wilder, 1865; reprinted Kuala Lumpur, Oxford University Press, 1965), p. 35.
31. Bogaars, "Tanjong Pagar Dock Company", pp. 133–35.
32. Quoted in Kratoska, *Honourable Intentions*, p. 49.
33. Makepeace et al., eds., *One Hundred Years of Singapore*, Vol. 2, p. 605. It was recorded here that 1,510 tons were loaded — slightly less than what the *Straits Times* had recorded. The high work output of the Chinese coolies in the Singapore port is not exceptional. In America, in the 1860s, the Chinese labored and set a record by completing more than 10 miles of track within twelve hours and forty minutes while laying the transcontinental railroad for the Central Pacific Corporation.
34. *Straits Times*, 5 August 1902. The world-record feat was also noted by Edwin A. Brown who was living in Singapore at that time. The happy occasion was somewhat marred by the death of one of the cruiser's workers whose funeral took place on board the ship itself just before she left the port. See Edwin A. Brown, *Indiscreet Memories: 1901 Singapore through the Eyes of a Colonial Englishman* (Singapore: Monsoon Books, 2007), p. 102.
35. K. Dharmasena, *The Port of Colombo* (Colombo, 1980), p. 41.
36. Quoted in Kratoska, *Honourable Intentions*, p. 50.
37. Henry Wilson, *A Naturalist in Western China: with Vasculum, Camera and Gun, 1913*, quoted in Alan Macfarlane and Iris Macfarlane, *The Empire of Tea* (Woodstock & New York: The Overlook Press, 2003), p. 103.
38. Ibid., p. 170.
39. *The Graphic*, 4 November 1876, quoted in *Straits Affairs*, edited by Tate, p. 12. In Chinese the term *k'u-li* literally means "hard strength". Foreigners residing in China often used the term to denote household menial Chinese labourers. However, from the mid-nineteenth century, the term took on a different coloration when European capitalists intensively recruited thousands of Chinese to meet the labour shortages on colonial plantations throughout the colonised lands. Even as far as the Chincha islands, off the southwest coast of Peru, Chinese coolies were indentured during the

mid-nineteenth century to work in the guano mines under severe working conditions. Each worker had a quota of five tons of guano per day. One American observer noted: "I seem to see them [the Chinese coolies] at their work their slender figures quivering under the weight of loads too heavy for them to wheel for everyone who went ashore remarked that they took loads altogether disproportioned to their apparent strength ... I observed coolies shoveling and wheeling as if for dear life and yet their backs were covered with great welts". Quoted in Evan Fraser and A. Rimas, *Empires of Food: Feast, Famine, and the Rise and Fall of Civilizations* (New York and London: Free Press, 2010), p. 133.

40. Mary Turnbull, *History of Singapore* (Singapore: Oxford University Press, 1989), p. 93.
41. Lee Tai To, ed., *Early Chinese Immigrant Societies: Case Studies from North America and British Southeast Asia* (Singapore: Heinemann Asia, 1988), p. 141.
42. Bogaars, "Tanjong Pagar Dock Company", p. 134.
43. *Singapore Daily Press*, 5 November 1877.
44. Frank Broeze, Peter Reeves and Kenneth McPherson, "Imperial Ports and the Modern Economy: The Case of the Indian Ocean", in *Port and Harbour Engineering*, edited by Jarvis, p. 377.
45. Brown, *Indiscreet Memories*, pp. 213 and 228.
46. *Annual Report of the Straits Settlements, 1910*, p. C238.
47. *Annual Report of the Straits Settlements, 1911 and 1914*, p. C116 and p. 20. After lengthy legal proceedings in the courts in London, the differences were settled in early 1914. The Dock Board paid John Aird & Co. a total of 664,000 pounds for all permanent work done and claimed possession of all the "contractor's plant and effects of every description and their stocks of materials".
48. *Annual Report of the Straits Settlements*, 1920, p. C 172.
49. *Singapore Free Press*, 2 January 1932.
50. B. Cable, *A Hundred Years of the Peninsula and Orient Steam Navigation Company, 1837–1937* (London, 1937).
51. The *Lady Mary Wood* was used to ferry troops from India to control the 1848 rebellion in Ceylon. P&O ships also took part in freighting opium to China and taking silk out.
52. *Straits Times*, 5 August 1845. One mail related that "The Earl of Aberdeen moved in the House of Lords an address to the Queen expressing the pleasure which the House concurred with Her Majesty in conferring on Sir Henry Pottinger a pension of 1,500 pounds per annum".
53. A. Wright and H.A. Cartwright, *Twentieth Century Impressions of British Malaya: Its History, People, Commerce, Industries, and Resources* (London: 1908; reprinted Singapore: Graham Brash, 1989), p. 42.

54. *Proceedings of the Straits Settlements Legislative Council, 1907*, C188.
55. For a study of the impact of the transition from sail to steam shipping on coastal shipping in Britain, see Mark Freeman, Robin Pearson and James Taylor, "Technological change and the governance of joint-stock enterprise in the early nineteenth century: The case of coastal shipping", *Business History* 49 (2007): 573–94.
56. Despite the lack of primary sources which were destroyed during the Second World Wars, the celebrated history of Mansfields and its associated company Straits Steamship is succinctly documented by Eric Jennings. See Eric Jennings, *Mansfields* (Singapore: Meridian Communications, 1973).
57. Ibid., pp. 22–23.
58. These collaborations between the Chineses compradors and the Western enterprises often allowed the more astute local businessmen to amass considerable wealth or capital to finance their own business. In Hong Kong, for example, Kwok Acheong who worked for the Peninsula and Oriental Line, had himself become a significant steamship operator.
59. William Laxon, *The Straits Steamship Fleets* (Sarawak: The Sarawak Steamship Company Berhad, 2004), p. 7.
60. Jennings, *Mansfield*.
61. The pride of all the company's fleet was probably *Kedah*, which at 21 knots outran the Malay Express railway. It had a celebrated history. Launched by the Lady Maxwell on 16 July 1927, the steamer became very popular and was nicknamed "the little queen of Malacca Straits". She was a white, impressive looking ship, with mahogany-coloured life boats. Amid ships she had accommodation for about eighty first-class passengers. Shortly before the fall of Singapore to the Japanese, *Kedah* was rushed to assist in the general evacuation. She made several voyages and returned to take refugees and survivors. When the British forces of liberation arrived off Singapore on 5 September 1945, HMS *Kedah* was in the vanguard with the Admiral on board. For interesting first-hand accounts by sea captains and river steamer captains of the Straits Steamship, see Charles Allen, *Plain Tales from the British Empire* (London: Abacus, 2008), pp. 631–50.
62. Jan Morris, *Pax Britannica: The Climax of an Empire* (Penguin Books, 1981), p. 52.
63. Quoted in Kratoska, *Honourable Intentions*, p. 44.
64. Keith Robbins, *The Eclipse of a Great Power: Great Britain 1870–1975* (London: Longman, 1983), p. 38.
65. Ibid.
66. Quoted in Kratoska, *Honourable Intentions*, p. 80.
67. B. Semmel, *Liberalism and Naval Strategy: Ideology, Interest and Sea Power during the Pax Britainnica* (Boston: Allen & Unwin, 1986), p. 52.

68. Donald Read, *Cobden and Bright: A Victorian Political Partnership* (London: Edward Arnold, 1967), pp. 103–16.
69. Quoted in Semmel, *Liberalism and Naval Strategy*, p. 51.
70. For a detailed examination of the interconnections between Britain's commercial expansion and her rise to maritime supremacy, see Paul Kennedy, *The Rise and Fall of British Naval Mastery* (London: Fontana Press, 1991).
71. *The Times*, 14 February 1938.
72. Northcote Parkinson, *Britain in the Far East: The Singapore Naval Base* (Singapore: Donald Moore, 1955), p. 28.
73. Ibid.
74. Ibid.
75. *The Times*, 15 February 1938.
76. John Dower, *Japan in War and Peace: Essays on History, Race and Culture* (London: Harper Collins, 1993), p. 265.
77. Quoted in Nicholas Baker, *Human Smoke: The Beginnings of World War II, the End of Civilization* (New York: Simon & Schuster, 2008), p. 183.
78. For a detailed account on the role of the Naval Base in the British defence policy of Singapore and Malaya from 1918 to 1941, see Ong Chit Chung, *Operation Matador: Britain's War Plans against the Japanese 1918–1941* (Singapore: Times Academic Press, 1997).

4
Introducing Technological Systems

Between 1870 and 1930, the trading port of Singapore changed in ways that Stamford Raffles could never have dreamed possible. The opening of the technological wonder in the form of the Suez Canal in 1869 heralded an exciting era for the trading town whose population of less than 100,000 depended largely on the buying and selling of local and imported goods. The healthy economy had attracted ever-increasing numbers of immigrants during the period 1870 to 1900. The population increased from 97,111 in 1871 to 228,555 in 1901, largely due to a massive arrival of about 110,000 Chinese during this period.[1] The number of Europeans and Americans also showed an increase from 922 (of which 594 were British) in 1871 to 2,861 (of which 1,880 were British) in 1901.[2] Together with other prominent colonial cities like Algiers, Cairo and Delhi, Singapore was developed to serve as an economic and political centre of British rule in this region.

With the arrival of the steamship into the expanding harbour, the people slowly but surely began to enjoy the fruits of Western industrialization which took place thousands of miles away. Ships brought along goods, machinery, equipment and household gadgets of all kinds — and foreigners of different races and with different skills and professions. In the United States and Britain, by the end of the nineteenth century, the process of industrialization led to the rapid transition of pre-industrial rural towns to industrial urbanized cities. In the quest for urban environmental improvements in the modern era, medical doctors prescribed new standards, engineers constructed more effective equipment and civil servants created new public institutions. While American engineers built with anticipation of advances still to come, British engineers and technologists often indulged in what Arnold Pacey termed as "idealistic engineering" — constructing huge monuments to idealize "Progress", such as in Britain where "engineers working on municipal sewerage and water projects or employed by the Metropolitan Board of Works in London built monumental pumping stations and other structures of extravagantly high quality".[3] Britain's colonial "estate" like Singapore was gradually transformed, especially from the last quarter of the nineteenth century, by technological systems that stimulated individuals to become more dependent on one another. However, before the transfer of colonial administration to the Crown Office in 1867, projects in public works and health services under the East India Company were largely uncoordinated and local administrators had to persuade the Company's Directors of the benefits of such public works for a trading company.

LINKING SINGAPORE WITH THE WORLD

In 1854, Governor-General of India Lord Dalhousie wrote to a friend: "The post takes ten days between [Calcutta and Bombay]. Thus in less than one day the Government made communications which … would have occupied a whole month — what a political reinforcement!" British imperialists, like Dalhousie, were simply enthralled by the development of the telegraph lines. As early as 1816, the telegraph invention (by Francis Ronalds) was first offered to the British Admiralty to enhance seaborne communications. But it was brushed aside and the wonders of telegraph were instead exploited by the private sector in Britain. By the 1840s it was clear that the telegraph would revolutionize overland communications. Telegraph technology played a significant part in suppressing the Indian Mutiny of 1857 in which one mutineer, on his way to execution, picked on the telegraph as "the accursed string that strangles me".[4] It also served, for the first time, to educate the public of war outside the comforts of their home. As explained by historian Donald Cardwell, the telegraph "played a part in the Crimean War; for the first time in a major war newspaper correspondents and the telegraph played a part in keeping a critical public well informed of the conditions under which the war was fought and its disasters were duly reported".[5] The crucial technical change which contributed greatly to firmer control and rapid expansion of the British Empire was the development of durable undersea cables. Gutta-percha, an imperial product from the colony of Singapore and Malaya, made this possible and the first cross-Channel cable was laid in 1851. On 27 July 1866, the Anglo-American Telegraph Company's cable finally reached the American coast. It marked the dawn of a new era.

The ability to send a telegram between London and New York in 1866 and receive a reply within an hour must indeed have seemed a wonder. A letter and its reply sent in the conventional manner, by steamship, would have taken three weeks. And this astonishing feat was accomplished a mere thirty years after the first electric telegraph was demonstrated. The speed of scientific and technological advances during the early years of Queen Victoria's reign was quite staggering. When the Atlantic cable was completed in 1866, the speed and frequency of communication between nations across the great seaboard increased, thereby permanently changing the character of diplomatic negotiations. The telegraph cable laid the foundation for the growth of international trade (particularly the growth of multinational corporations) in the later decades of the twentieth century. The success of the Atlantic telegraph cable led to the laying of other more ambitious submarine cables, many of which were used to link London with the British Empire. By 1880 there were altogether 97,568 miles of cable across the world's oceans, linking Britain to India, Canada, the Straits Settlements and Australia.

The telegraph as a communication technology was transferred to colonies like India, Singapore and Malaya with a political purpose — to maintain the power and security of the British Empire. Rapid means of communication enabled politicians in London to monitor their agents in the far-flung corners of the world, and the advent of the telegraph made all this possible. The land and submarine cables were seen as symbolic tools to realize Britain's majestic dream of an all-British world trading system. Rudyard Kipling's "The Deep-Sea Cables" echoes this dream:

> They have wakened the timeless Things; they have killed their father Time;
> Joining hands in the gloom, a league from the last of the sun.
> Hush! Men talk today o'er the waste of the ultimate slime,
> And a new Word runs between: whispering,
> "Let us be one!"[6]

The paramount need to protect British India meant an early exposure for Singapore to the wonders of electric telegraph. As early as 1827, the subject of telegraphy was surfaced in the official circle. In that year, the Inspector-General, recognizing the need to establish telegraphic communication, was requested to submit a proposal and budget for the construction of three telegraph stations and a lighthouse. The document soon arrived in the Court of Directors of the East India Office:

> The Court of Directors appeared to have been astounded at the audacity of the telegraphic proposal. In a despatch dated June 17, 1829, they wrote: "You will probably not find it expedient to erect at present the proposed lighthouse at Singapore, and we positively interdict you from acting upon the projected plan for telegraphic communication. We can conceive no rational use for the establishment of telegraphs in such a situation as that of Singapore!"[7]

It was yet another indication of the East India Company's indifferent attitude towards its commitment and control of Penang, Malacca and Singapore, which, in 1826, were collectively known as the Straits Settlements. Especially after 1830 when the Company lost its monopoly of the China trade (which the Straits Settlements had been acquired to protect and supply), the Directors adhered strictly to a negative policy towards the Straits, trying to avoid financial deficit and avoiding all unnecessary relations with the Malay States in the Peninsula by a rigid policy of non-intervention.

By 1850, it was clear to merchants and government officials in Singapore that the island was a thriving entrepôt port and that effective business communications with the outside world was of paramount importance. Technology, in the form of the telegraph, was the key. In 1859, with financial support coming from the Australian colonies and the Dutch government, a submarine cable linking Batavia to the island was laid. Congratulatory messages were relayed on 24 November of the same year; but there were "frequent interruptions of communication, owing to the cable having been broken, either from friction against the coral reefs over which it passes, or to its having been dragged by the anchors of vessels anchoring in the narrow straits in its line of passage".[8] The cable was repaired on a few occasions but the reception continued to be so poor that the line had to be abandoned. Prior to 1860, underwater cable technology was still in its experimental stage. The cable was too thin, consisting of a few strands of copper wire coated with gutta-percha and wound with hemp. The failure of the telegraph line between Singapore and Batavia

Introducing Technological Systems

heightened the urgent need to link the former with India by overland cables. It was suggested that "a more simple and less expensive telegraph might be carried overland from Singapore to Rangoon, the latter being already in communication with India".[9] All this time, however, the Home Government in Calcutta "resolutely declined to assist, and though repeated deputations waited upon it on the subject, it refused to alter its policy".[10] It was left to Lord Canning, the Governor-General of India, to acknowledge in 1859 that the shortcomings of the Indian Office bureaucracy were the "greatest evil" in the development of the Straits Settlements.[11]

Singapore was eventually plugged into the British imperial communication network in 1869 when permission was given to the British Australian Telegraph Company Limited to lay a cable between Singapore, Java and Australia. Some of the public buildings that were connected by the telegraphs in 1873 included the Government House, the Public Buildings and the Chief Police stations.[12] The company was later incorporated into the Eastern Extension Australia and China Telegraph Company Limited.[13] The Eastern Extension was an associate company of Britain's great cable company, the Eastern Telegraph Company. It opened its first office in a building in Prince Street in 1870, before moving to Raffles Quay in November 1895. As Singapore progressed, the office accommodation was found inadequate to meet the ever-increasing telegraphic business of the city. In July, the company moved again to its final home at Robinson Road where, before the war, some 10,000 cablegrams were transacted daily. Meanwhile, sea cable technology was gaining good progress. More sensitive sending and receiving instruments were developed and laying techniques at the right spots on ocean beds were improved. In 1879 the siphon recorder, invented by William Thomson, visually displayed messages sent. The capacity of the cable was doubled when the technique of duplexing, which allowed simultaneous transmission in both directions, was introduced between 1875 and 1879. Finally, before the century closed, the automatic transmitter further eased the operations of telegraph clerks.[14]

By 1882, submarine cables had connected Singapore to the world. The telegraph system was the very first network of technological systems that the people in Singapore experienced. It had become crucial to the economic and political life of the colony. The communication technology was a great boon to the business community especially since Singapore was developing into an important shipping junction. It enabled merchants and shipping agencies to maintain speedy and reliable contacts with ports and the world market at large. Newspapers, too, had become dependent on the telegraph for quick transmission of important information. Indeed, submarine cables played a significant during the First World War. Of particular interest to the people in Singapore was that, just after the elusive German raider *Emden* had made a daring raid in Penang, she was destroyed when attacking the cable station at Cocos Islands. Her conqueror, H.M.S. *Sydney*, was summoned by a wireless S.O.S. from a cable operator at Cocos.

Besides the laying of submarine cables, significant progress in intracolonial telegraph network was also achieved in the Straits Settlements after 1885. Unlike submarine cables, these land lines were neither long nor expensive. The rapid extension of the telegraph lines in the Straits Settlements, together with a few of the dependent colonies, is shown in Table 4.1.

The statistics revealed two important trends. First, the figures for the Straits Settlements implied a healthy economy since expansion of telegraphic services was a direct response to the needs of the business community, the single, largest user of the facility. Likewise, this also explained, with local variations, the progress of telegraph operations in the other colonies of the empire. Second, the figures gave an indication of the nature of the policy of the British government towards the extension of telegraphic services as a form of control of the wide empire. By the 1870s strategic factors, in terms of ensuring full communications with British India, were overwhelmingly important in determining the laying and

Table 4.1
Government Telegraph Operations
(In miles of Operating Lines)

Colony/Year	1885	1890	1895	1900	1904	1908
Straits Settlements	130	635	1,200	1,322	1,579	1,654
Ceylon	676	794	997	1,438	1,611	1,754
Mauritius	103	122	135	135	293	331
Gold Coast	109	177	414	688	1,032	1,348
Jamaica	542	637	820	828	880	900
British Guiana	260	260	272	311	324	324

Source: Great Britain, Parliamentary Papers, "Statistical Abstracts: Colonies", *CVI* (1910): 280–81, cited in Richard Kesner, *Economic Control and Colonial Development* (Oxford: Clio Press, 1981), p. 139.

control of telegraph cables in the Empire.[15] Recognizing how telegraph networks could enhance imperial defence, the Admiralty and War Office advocated an "All-Red" or "All-British" cable system, that is, a cable network which linked all parts of the Empire without ever touching foreign soil. It required every colony or naval base in the British Empire to possess at least one cable linked to British territory; in another word, military consideration had priority over commercial needs when laying the initial cables. The Eastern Telegraph Company's line in the East was, from the very beginning, of the greatest strategic importance and it was the first link of the "All-Red" system. From India, extensions were laid to Singapore, and thence to Hong Kong; and from Singapore a branch line was laid to Australia and New Zealand.[16] Singapore again came into the grand defence strategy when it was decided in 1901 to add a spur to the new and extensive link between the Cape and Australia by the Cocos-Singapore line. It was a decision to give further protection to the cable communications between Britain and India.[17] Thus, Singapore's own strategic position made her very much a part of the British network of imperial defence.

In times of international disputes, early understanding of the situation, especially if it took place at the periphery of the Empire, was of utmost importance. As an innovation in communication technology, the telegraph could allow the Colonial Office to gather information and address disputes swiftly and effectively. However, this was not quite the case at this time because of the initial high cost and the length of time taken to deliver and decode messages. As late as 1890s, the Colonial Office still depended on despatched mails from fast steamers rather than telegrams.[18] Hence, the British intervention in the Malay States seemed to indicate a lack of communication and control despite the presence of the telegraph. In 1874, two years after telegraphy was introduced to Singapore, conditions in Perak became so chaotic that Lord Kimberley, the Colonial Secretary, instructed Sir Andrew Clark, the Governor of the Straits Settlements, to "ascertain and report" on the matter.[19] But the British government was committed to something more drastic than expected when Clark pushed for the signing of the Pangkor Treaty in 1874 which launched the Residential System of indirect rule. These momentous events occurred without prior consultation with London, in spite of the newly laid telegraph cable. The implication was that the man-on-the-spot, such as Andrew Clark, was given the benefit of doubt by the Colonial Office since the latter was not able to have an accurate picture of the crisis in a country far away from the centre of administration. The man-on-the spot knew best of the local conditions and the government had to support his decision. Indeed, Clark's action probably set a precedent for other men on the spot in the colonies to proceed with their decisions without telegraphing London to seek permission. Telegraphy technology failed to stop those imperial representatives from many making decisions which affected the course of history for the many former colonies of Britain. If anything, it forced the Colonial Office to support rather than curb the actions of local officials.

Another milestone in Singapore's telegraphic communication history was reached in 1929 when the colony was linked directly to London. Before this, cablegrams to and from England were relayed from station to station along the "imperial" cable route, each station representing a check in the speed of the message. With the introduction of the Regenerator system in early 1929, message signals were automatically strengthened at each of the eleven stations and relayed without any pause whatever between England and Malaya. The total mileage of 10,521 miles was bridged in one to two seconds, as compared to ten to twelve hours in 1870.[20] In the meantime, further technical progress was made in Singapore and Malaya to provide a more efficient telegraphic service. Due to the long distances between major towns, a combination of Wheatstone and Morse telegraphic systems were implemented because they prevented interruptions and prepared for high speed of transmission.[21]

Associated with the introduction of the telegraphic cable into colonial Singapore is the honour of being the first city in the East at this time to have a telephone system. It happened in 1879 — just three years after Alexander Graham Bell had filed a patent for one of the world's greatest inventions — when the telephone was introduced to link the trading port at Tanjong Pagar to the commercial hub at Raffles Square. And for the next seventy years, the telephone service was supplied and operated by private enterprise. The move was made by Bennett Pell, the manager of the Eastern Extension Telegraph Company.[22] He started Singapore's first private telephone exchange in one of the offices of Paterson Simon & Company in 1879 by attaching its lines to the government telegraph poles lining the streets of the commercial area. Although it was plagued with initial technical problems, there was no doubt that the telephone had arrived and made its mark. Pell's venture was soon bought over by the Oriental Telephone & Electric Company (OTEC), a British company registered in London in 1881 for the purpose of developing telephone patents of Bell & Edison. OTEC opened its Singapore branch, situated in Robinson Road, on 1 July 1882. By 1894 the number of operating telephone lines had increased to 256 and by the turn of the century, the telephone had become an integrated aspect of residential and business life in Singapore. To meet the rising demand for telephone service, technical and infrastructural changes took place rapidly. Between 1904 and 1907, an extensive underground cable system was laid and a new telephone exchange was built at Hill Street. After the war in 1920, a battery system was installed and equipped to handle 5,000 lines. A decade later, this central battery system was replaced by an automatic, step-by-step strowger switching system. The technological change allowed the Police authorities in Singapore to install its own automatic telephone network but still linked to the central system. It was hoped that "this branch exchange will facilitate communication between the public and the police as a sorting office for messages received".[23] Further technological improvements came in the form of the use of teleprinter system in all main telegraph circuits which were designed to eventually replace the Morse system. Telephone trunk communication between Singapore and Kuala Lumpur was opened on 1 August 1931. OTEC continued to be the only private supplier of telephone lines and services in Singapore, and employed 300 staff including nine Europeans in 1931 to operate 5,000 lines.[24] It was also not long that the inevitable link with the metropole of the British Empire, London, had to take place. On 1 December 1937, the long-awaited telephone service between Singapore and London opened and the first call, lasting three minutes, was made by Rosalind Wong, daughter of S.Q. Wong (Chairman of the Malayan Tribune Company) and a guest reporter of the *Daily Press* in London.

The telegraph and the telephone system created a closely connected business network of traders. However, the social change attributed to these communication innovations was due not just to the space-transcending and other technical capacities of the instrument and its support systems. It also depended on the easily mastered requisites of telegraphy and telephone use, the market character of the Singapore entrepôt economy and the willingness to use the technological systems in lieu of or supplement face-to-face interactions. British and Chinese merchants and traders recognized that the telegraph and telephone had effectively destroyed the end of traditional commodity markets where prices of commodities would vary from town to town within the same geographical locality and agents would buy cheap and sell dear by moving their commodities around in space. The new communication technologies now put everyone in the same place for the purposes of trade. It made geography less important and, simultaneously, it shifted speculation from space to time. With the spatial uncertainty of prices eliminated, agents began to buy cheap and sell dear by trading "futures" — as in rubber and tin industry of Malaya. Rubber and tin from the various plantations and mines had to be graded in advance to lend themselves to future trading and this required that the actual products were mixed or adulterated in order to achieve an abstract standard.

TECHNOLOGY AND TRANSPORTATION

While one reads of trams and motorcars plying the roads in colonial Singapore, most people still did most of their moving about on foot — to the market, work and other places in the vicinity of the town. Singapore was a "walking city". In May 1821, Singapore had about fifteen miles of road, "bordered by canals, in which flow free streams of fresh water, or by dry ditches" and, by the mid-nineteenth century, the roads were still targets of public complaints in the local newspapers.[25] Granite metal was not used and instead "all the roads require is, from time to time, a dressing of the laterite gravel which was dug from Public Works Department quarries by the roadside".[26] However, the monsoon seasons often wrecked havoc on the "First Class" roads which became "soft" and, with the heavy bullock cart traffic which passed over them, cut up. The roads were "reduced for a short time to the conditions of an English country lane".[27] While there was no direct reference in the annual reports of the Straits Settlements on the road conditions in Singapore in the late 1890s, it could be surmised from such reports on Malacca and Penang that the causes for the "deplorable conditions" of the roads in the latter were also applicable to the island colony. As explained by the Acting Superintendent of Works and Surveys, "want of money, bad weather and want of qualified supervision" were contributing factors.[28] The last factor was significant as the lack of technical expertise hampered improvements. As stated in one of the reports:

> The Public Works Department, however, suffers, like other Departments, by having only just the number of qualified Civil Engineers required to fill the places where their supervision is essential. It follows, as a matter

of course, that the Department is always short-handed. Every European officer spends, on the average, fifteen months in every seven years on leave... The Department is a spending one and it is false economy to grudge it sufficient skilled supervising officers. Matters are not as they are in Europe where competent engineers can be engaged for a few months whenever required.[29]

At the turn of the century, the majority of the population in Singapore — the 164,000 Chinese or nearly three-quarters of the total population — settled in and near the town centre. Before the introduction of motor transport in 1896, and even after that, the mode of transportation in Singapore experienced few changes. Bullock-carts, hand-carts, horse-drawn carriages and bicycles were the main means of moving goods and people. The conveyance of goods to and from the port and city was monopolized by bullock-carts. Though cheap, the system was liable to break down because of cattle plague, thefts of merchandise from the carts and even strikes by the cart drivers.[30] In the newspapers, there were many advertisements for fodder, harness and other requisites for maintaining horse-drawn vehicles. As in Kuala Lumpur in the late nineteenth century, the pony-trap was the main travelling conveyance for those who could afford it. For the many European businesses and Chinese merchants owning provision shops throughout the business district and Chinatown respectively, the bicycle was an important technological import. It allowed the Europeans the travelling mobility and the Chinese workers to pick up goods from godowns and to make home deliveries of daily and household necessities.[31] The Raleigh bicycle became a household name and was imported into Singapore by Robinson & Company.[32]

The idea of a railway linking to the port to other parts of the small island was mooted as early as 1865. But after nearly a decade of bickering between the government and the companies at the port as to its usefulness, ownership, and capital expenditure, the whole scheme was dropped. Practicality and economic rationality was the main explanation: "Owing to the small area of the three Settlements and their geographical position along the shores of the Straits of Malacca, the necessity of railways has not hitherto been felt, and, with the exception of a few miles of tramway in Singapore and Penang, nothing has yet been done in this direction, although considerable progress has been made in the Protected States of the West Coast [of Malaya]".[33] In any case, unlike the Malay Peninsula, the economic and social impact of the introduction of the "iron-horse" into Singapore was minimal.[34] The end of the "Railway Saga" was followed by the tramway episode. The use of steam engines in road transportation was transferred from Britain to Singapore when, in 1886, the first steam tramway service commenced. It was run by a private company, the Singapore Tramway Company, and dock companies were included in its management. The lines ran from New Harbour to the business centre at Collyer Quay. In 1905 the trams switched to electricity. Operation of the tramway, however, proved unprofitable because of the keen competition from man-pulled jinrickishas.

Jinrickishas were first imported from Shanghai in 1880 and immediately gained popularity as a cheap and convenient means of travelling.[35] The number of rickshaw coolies soon swelled to 5,046 in 1888.[36] The rickshaw coolie became a distinctive feature of Singapore's street life.[37] According to Edwin Brown who came to Singapore in 1901, "seventy-five percent of the Europeans used the rickshaws to get back and forth to the office, and for the Eurasians and other portions of the populace rickshaws were almost the only means of transport".[38] Particularly for the *tuan besar* (rich and important individuals), a runner would be deployed to assist in pushing the rickshaw up steeper roads. While the jinrickisha offered the locals a form of transportation within the commercial area, the contraption and the pullers were becoming a social problem towards the end of the nineteenth century. In order to regulate and control the issue of licences and the number of jinrickishas on the congested roads of Singapore, the Jinrickisha Ordinance was revised in 1899 and which now contained stringent measures (such as the imposing of heavy fines and suspension of licences) for malpractices by pullers and owners alike.[39] On 21 October 1901, as witnessed by resident Edwin Brown, the riskshaw pullers staged an islandwide strike:

> The origin of the strike seems to have been a demand from the police that the men who pulled rickshaws should know something of the rules of the road … and when the police made their reasonable demand it was immediately concluded by the ignorant men that the governor was out for blood.… The strike spread to the gharry-wallahs who were frightened to go on the streets for fear of being attacked, and as the days wore on, the situation worsened. Words went round that the Europeans would be advised to arm themselves, too …. and in Chinatown the police had to

Introducing Technological Systems

make arrests and shots were fired in the air to frighten to crowds. Along Rochor Road and in the district round the rickshaw station, the strikers were active and considerable damage was done.[40]

These strikes, however, did not prevent the rapid increase of jinrickishas plying the busy streets within the city area. In January 1904, single-seated rickshaws with rubber tyres were imported into Singapore. Up to this time the rickshaws had been iron-tyred and all doubled-seated. As at June 1907, there were 7,469 and by 1920 there were about 9,000 jinrickishas being pulled by an army of 20,000 coolies. The simplicity of the wheel and pure human labour had provided the masses with a cheap system of transportation. In the process, the jinrickisha was a bane to the tramway. Its mobility allowed for travel to nooks and corners of the city. The presence of the tramway reinforced, rather than hindered, the popularity of the two-wheel gadget, as more short journeys were undertaken by passengers after leaving the tram. Moreover and particularly for the Europeans, travelling in the tramcars was not all that comfortable because their conditions were described as "rather shabby and very noisy [and] drivers and conductors are also of an inferior class".[41]

While the masses had the jinrickishas to make short trips, the well-to-do were making a gradual switch from travelling in palankeen carriages to the motor car. The number of motor cars increased from 842 in 1915 to 4,456 in 1925. In 1926, Cycle & Carriage was incorporated in Singapore and, in the late 1920s, introduced the "Whippet Roadster" to the well-to-do at $1,875.[42] Most of the roads, however, were planned when less than half the volume of traffic was in existence and were not suitable to handle motorized traffic. The Municipality therefore carried out a systematic overhaul of roads in Singapore, widening and strengthening the roads to a standard width of sixteen feet of metalled surface composed of three layers of hand-packed granite, limestone and bituminous products.[43] But before this utilization of up-to-date technology and materials, road construction and the quality of roads in Singapore (and in Malacca and Penang) was often the target of criticism. Nevertheless, road construction continued to improve and by the end of the First World War, many of Singapore's roads were converted from laterite to granite. By 1920, 96.26 miles of metalled road provided easy mobility to the people living in the congested island.[44] Travelling between places in British Malaya, as a whole, was now faster and safer as the country was linked with a network of roads and railways. It was a far cry from the days of Isabella Bird and Emily Innes. Technology had shortened space and time and, as noted by the British official Eric Macfadyen, it even brought ice — a product of modernity — to the villages in Malaya.[45]

In the late 1930s, improved technology soon made road breaking a much easier task when the Municipality imported machines like the Warsop Road Breakers and Rock Drills.[46] The construction method which proved the best was the use of bitumen to bind roads. Bitumen tar or asphalt surfaces which could bear up under the weight of increasing motorized road traffic were placed over macadam slag or gravel bases. Although improvements in road transport and road-building technologies gave the Europeans and masses convenience and ease of travelling, the dramatic increase in various types of motor vehicles caused serious congestions in the city, leading to complaints of adverse psychological effects because of "inconsiderate sounding of motor hones".[47] Writing in 1926, one contemporary observer commented:

> Singapore has as many noises as most places, and being a large town in the tropics, the noises are continuous and varied, both by night and day. During the day, prevailing over all will be the hooting of the motors. Sometimes this is almost deafening, and much of it is due to the Chinese small-bus drivers, who can only get through the traffic by blasting his way through it! ... In addition, the rickshaw-puller never bothers to get out of the way unless he his driven out by superior force or by a great noise. You will, then, hear motor-cars on all sides by day and by night.[48]

Many observers who came for a revisit after some years of absence invariably commented that the greatest changes took place on the roads and streets of Singapore:

> A decade ago (1925) it was said that traffic was made up of mosquito buses, rickshaws, bullock carts, gharries, and trams; police control was difficult because there were no pavements and pedestrians swarmed. Every vehicle went at its own pace. Today [1935] omnibuses and trolley buses have superseded small buses in the main parts of the city. A bullock cart is rarely seen; there are less than a couple of dozen gharries left. Plenty of rickshaws, but their numbers are gradually dwindling. Singapore street traffic now consisted of swiftly-moving vehicles controlled by one of the most efficient traffic police organisations in the world... Automatic signal

lights similar to those in London are likely to be installed at congested crossings in the near future.[49]

In the late 1930s, electrification further enhanced the rapid urbanization of the Singapore city.

THE ELECTRIC SYSTEM

The development of electrification was a continuous process of fruitful application of scientific and engineering expertise in steady technical innovation of increasing complexity since the early decades of the nineteenth century, and culminating in Thomas Alva Edison's remarkable construction of the famous Pearl Street Station in the heart of New York City over the period 1880 to 1882. During this time, Edison applied for 156 patents. National legislations in the United States and Britain and favourable economic conditions before 1914 also determined the rate and nature of electrical development. While the adoption of electric power in major industries, such as shipbuilding, steel, mining, textile and engineering, in the United States and Britain gained quick acceptance by the beginning of the twentieth century, the spread of electricity for domestic purposes was, even by the 1920s, slow and mainly an urban phenomenon. It was too expensive for the ordinary household where gas or coal systems were the main and affordable energy source.

For most colonies and dominions located at the "periphery" of the British Empire, the wonders and convenience of the electric system was something which the masses got to enjoy way after it had lighted up the rural towns and cities of the industrialized West. Before its arrival, the main streets in the town of Singapore were first lighted with oil lamps in 1824. Even with the introduction of gas lighting on 24 May 1864 — to celebrate the birthday of Queen Victoria — the Singapore town plunged into darkness at nightfall. The very brightest gas street lamps provided less light than a modern 25-watt bulb. Used in the house, gas lighting often blackened ceilings, discoloured fabrics and corroded metals. Lighting a gas jet carelessly could also start a fire. However, within the confines of the house, gas produced the brightness, at least compared with anything else the pre-electric world knew. Consequently, it led to changes in the pattern of home activities. It made reading, card-playing and even conversation more agreeable. Gas lighting came about in 1860 when a private company,

Singapore Gas Company, constructed and supplied gas at the gasworks situated in Kallang. It was reported that "[w]hen the lamps were lighted, natives were seen going up to the lamp-posts, and touching them gingerly at first with the tips of their fingers; they could not understand how a fire could come up at the top, without the post getting hot, which was by no means unreasonable, as they could not know what the gas was".[50] But it was a welcoming sight as the use of modern science brought yet another feature of the "civilizing" influence and power of industrialized Britain. The Chinese elites, like Whampoa, also wanted to be part of modernity and clamoured for the supply of gas to their households. It was only with the purchase of the gasworks by the Municipality in November 1901 — in response to constant public complaint of inefficient services by the privately owned gasworks — that the gas lighting system was made easily available. But it remained a middle-class indulgence; the poor could not afford it and continued with the use of oil-lamps or candles.

Although electric street lighting, purchased in bulk, was provided in 1906 to the town's central area, as late as the 1920s, electrification of the Singapore city "was a rarity and mostly supplied at great expense from private plants".[51] Hence only the business districts and European homes enjoyed such luxury. A request was also made in 1893 to have the Government House lighted by electricity, and the high cost was partly to be paid by the Governor and his successors.[52] In response to an increasing interest in electrification, the local authority commissioned two separate reports to look into the viability of electric lighting. The first report was submitted by electrical engineers H. Burstall and Edward Monkhouse in 1899 and it highlighted that "[f]or an ordinary dwelling home occupied by a family using say 40 lamps this would work out to about $30 a month — a sum which comparatively few Singapore householders could afford to pay".[53] A year later, another report also stressed the high cost involved. Municipal engineer Corbet Woodall reported that "the population likely to require electric lighting is not concentrated within a small area but is grouped in places which are at considerable distance apart [and] a general supply over the whole area would be very costly, so much so that the town could hardly face the expenditure in years to come".[54] The abundant supply of imported coal could also have delayed the legislative process of supplying electricity to households. The invention of the incandescent filament lamp, especially Edison's high-resistance variety, was crucial here. For the first time, electricity offered something useful not only in industry and commerce, but in every home. It also led to the demand for a centralized system of power generation and distribution.[55]

Although the Municipality built the first power station, the St James Power Station, in 1927, many did not enjoy the wonders of electricity even as late as 1930s. Even in London, as late as the 1930s, almost half of the city's streets were still lit by gas.[56] As Singapore's streets and roads became congested with vehicles of all kinds, there were frequent complaints by the public of the quality of road lighting on main traffic routes. When accident cases were heard in the court, the accused often pleaded poor lighting as the cause. In 1938 a committee was asked to investigate on the lighting situation and it came up with interesting comments:

> Proper illumination of streets in Singapore raises problems which do not exist in England, for example, the presence of jinrickshas and the number of pedestrians with dark clothing who through the lack of sidewalks are compelled to use the carriageway … It could therefore be argued that Singapore requires an even higher standard of lighting than that found in other populous cities…. We have satisfied ourselves from inspection of many streets that the existing system of lighting on traffic routes by gas is not satisfactory. Gas standards do not comply with modern requirements as to the height of the lantern required to give a proper distribution of light with absence of glare and the number of lumens, or units of illumination. We consider electric lighting should replace the present gas lantern on all main traffic routes.[57]

Besides lighting up the Singapore urban town in the 1930s, electricity also allowed the introduction of another modern technological wonder — the air-conditioner. News of the use of air-conditioning was reported in the *Singapore Free Press* in October 1936 when businessman Lester Goodman fitted his office with air-conditioning equipment.[58] However, the press lamented that "[i]n a rich city like Singapore where the burden of the heart of the day is so heavy, it is strange that air-conditioning has been so long in arriving, and even yet it is only a curiosity, with no prospect of it becoming general for many years [and] there are possibly two reasons for this, one is the traditional tidapathy associated with Malaya and the other is the expense".[59] Although the newspapers at this time frequently carried the advertisement by United Engineers Limited of the sales of "Carrier" air-conditioners and services provided by trained engineers, the air-conditioner was a luxury for most people.[60]

CIVIL ENGINEERING

In the 1820s and 1830s there were a multitude of wooden structures within the Singapore town. Skilled carpenters and artisans, mainly from Malacca, constructed public buildings that housed government offices, shophouses, tenements and other dwellings for the rapidly expanding population in the island. Chinatown itself was a "wooden" town. The danger was fire. On 9 February 1830, a great fire engulfed almost half of the wooden structures along Market Street, Circular Road and Phillips Street. To make matter worse, the fire-fighting brigade was manned by largely untrained convicts. However, the positive outcome was that the urban features of Chinatown was transformed with wooden structures giving way to brick buildings, the construction of which was supported by the mercantile community. By the last decade of the nineteenth century, the town of Singapore was visibly changed since Raffles first landed in 1819. But in the mid-nineteenth century, the Singapore town was in a shanty state. The *Straits Times* reported:

> Filth was thrown on the roads, gutters were uncleaned and everywhere there were pools of stagnant water. Carcases of dogs, cats and horses littered the roadsides, and the town was full of beggars and drunken sailors. Convicts and cattle were buried near the town and often so shallowly that dogs scraped away the earth and exposed the rotting corpses.[61]

From the last quarter of the nineteenth century, the Singapore town was beginning to take on a more imposing appearance. Several significant public works were completed by 1890. The Telok Ayer reclamation project (started in 1880) was largely completed at an estimated cost of $300,000. A large bay area was filled and "a wide extent of valuable building land secured in a direction where the town had become most congested [and] a direct road has been opened of the docks, and a steam tramway brings them in close connexion with the commercial quarters of Singapore".[62] Dredging and improvement works at the mouth of the Singapore River, as recommended by Sir J. Coode, were also completed and "with the mud from the river, the Esplanade has been doubled in width, much to the satisfaction of the European and Eurasian community, who find there the necessary accommodation for various means of recreation".[63] Another interesting project completed at this time was the development of the present area at Pulau Saigon. As described in the *Annual Report of the Straits Settlements* in 1890:

> Kampong Saigon is an island which was a few years ago covered with huts upon piles occupied by some of the worst characters in Singapore. They were cleared out, the island raised several feet with town refuse and mud from the river, and the whole has been surrounded by a neat river wall on a novel and economical plan. One portion has been given to the municipality for an abattoir, and that body have constructed one handsome iron bridge leading thereto…[64]

Finally, business at the waterfront of the commercial district was given a boost with the completion of the new Johnston's Pier. With the construction of a "wave screen, which is intended to secure smooth water alongside the landing pontoons", Johnston's Pier became a hub for the exchanging and landing of goods and passengers from surrounding islands.[65]

In the nineteenth century, brick, cut stone, plain concrete, cast and wrought iron, and wood were standard building materials. Practically all these had to be imported. Although there were several brick-making factories in Singapore, the local supply was insufficient to meet the needs of a rapidly urbanizing city. Bricks had to be imported from Hong Kong. The scarcity of bricks, coupled by the great demand for the material, led to general rise in the cost of construction in the early decades of the twentieth century. In 1844, bricks were selling at $18 per *laksa*; by 1902, they were priced at $50 to $140 a *laksa*.[66] Rising cost, however, was matched by technical advances in the construction industry which led to rapid urbanization of the Singapore town. From the late nineteenth century, technical advances in building construction led to a more widespread use of "Portland" cement and in combination with iron bars and girders.[67] This allowed the construction of brickwork structures which could be supported over wider openings. "Condor" cement was also introduced into the building construction during the early years of the twentieth century. It was imported by C. Dupire & Co. (name was later changed to Dupire Bros.). The material was randomly tested by Swan & Maclaren, and the engineering firm confirmed that its "tensile strain which it stood as neat cement or mixed with sand was considerably above that required for large engineering and architectural works".[68] By the 1930s, in addition to these materials, iron and steel supports, insulating boards, anti-corrosive paint, and reinforced concrete were widely adopted by civil engineers and builders. The introduction of fireproof tiles also lessened the fear that fire would buckle and collapse iron structures. "The vogue for modernism, which had found its way into England in the 1930s as a result of the work of avant-garde architects in Germany and elsewhere on the continent, came also to Singapore" and this knowledge and skills were transferred to local builders through printed journals and first-hand visits.[69] Standard new features, such as air-conditioning systems and electric lighting fixtures, were added and helped to "change the pre-war house from being essentially conservative in form and style into something more radical".[70]

As the population grew and wealth increased, the movement of the Europeans to the city-suburb was accelerated. Here wealthy merchants, ship owners, and government officials built attractive colonial villas and mansions. Colonial architecture reflected fashions in Britain but it was often adapted to incorporate local designs. Hence, the Classical or Gothic style from the 1840s was preferred, but buildings in Singapore and Malaya (and in India) often took on Oriental domes, towers and doorways. In the suburban districts of Tanglin and River Valley Road, the bungalow, with its distinctive thatched roof and veranda, was the popular architectural dwellings sought by the British middle-class. Thomas Metcalfe maintains that the bungalow "advanced a political purpose — that of social distancing because of its walls, sweeping drive, watchmen and gates".[71] But it was not quite a case of social distancing. The British residents in Singapore were constantly in touch with their servants who handled largely the day-to-day running of the household and in receiving local and foreign visitors at the veranda. In reality, the bungalow was just another symbolic image of the British practice of class-hierarchy and one which reflected the suburban attitudes of the middle classes. Similar to the role of the huge bungalows and villas, government buildings and modern hotels in Singapore were constructed on a grandiose scale, and designed to impress the subject peoples, represent power and convey a self-confidence and permanence that local, indigenous architecture sometimes did not. The two prominent hotels, constructed during the nineteenth century and which stood the test of time were the Hotel de l'Europe (in 1857) and the Raffles (in 1887). An amusing story was told by mining engineer Norman Cleaveland who stayed at the Raffles Hotel after journeying across the Pacific from America. He was shocked at the "extremely primitive" excrement disposal system known to locals as the *jamban* (Malay word for toilet) and the use of the "Shanghai jar". The receptacle used in the *jamban* was popularly known among Europeans out East as a "thunderbox". The "Shanghai jar" was a huge ceramic receptacle used by the locals to contain cold bathing water. Like many other Europeans who first arrived in this part of the world and seeing the huge jar for the first time, Cleaveland mistook the jar's purpose: "I thought you were supposed to get *in* it so, much to the distress of the servants, I got in and bathed in the Shanghai jar. I thought it looked a little tight but it was kind of cosy".[72]

Besides the construction of impressive government buildings to reflect imperial power and control, railway stations and bridges, often named after governors, were also potent symbols of British industrial strength.[73] Colonial Singapore did not have any massive, iconic buildings or structures that would attract world attention. The colonial administrators, however, were proud of the Victoria Memorial Hall, the magnificent neo-classical Municipal Building (later known as the City Hall), the Supreme Court, the illustrious Raffles Hotel, the majestic St James Power Station, St Andrew's Cathedral — and the bridges, such as Cavenagh Bridge, Anderson Bridge and Elgin Bridge, spanning the busy Singapore River. Elgin Bridge was completed in 1862 (rebuilt in 1926). Seven years later a cast-iron, suspension bridge was built by P. & W. MacLellan of Glasgow and named Edinburgh Bridge to commemorate the 1869 visit of the Duke of Edinburgh. This bridge was assembled by convict labourers — the last major project undertaken by convict labour in the colony of Singapore. Edinburgh Bridge was later renamed Cavenagh Bridge in honour of Governor William Orfeur Cavenagh, the last India-appointed Governor of Singapore, who decided to link the government offices in Empress Place with the commercial heartland of Raffles Place. In 1909, the bridge was converted to a footbridge with the opening of the Anderson Bridge, regarded as the most imposing of the bridges over the Singapore River, which ran parallel to it. The sign prohibiting bullock carts, horses and heavy vehicles from crossing it remains to this day.

By the 1930s, the emergence of the modern Singapore city as a bundle of large technological systems transferred from the industrialized West was seen as a testimony to Britain's desire to bestow modernity on her key imperial cities within the Empire. Within the rapidly changing urban environment in Singapore, British administrators also worked hard to prevent disease and promote public health — with the initial intention of maintaining and improving the quality of life of the European community living in the torrid tropical climate. However, the health of the masses could not be disregarded because Europeans and Asian immigrants alike lived in the shared environment. In any case, the evolution of public health from sanitary measures to environmental health and the wider concerns of social medicine progressed very slowly in the nineteenth century. Overcrowding coupled with *laissez faire* policies, lethargy and lack of will by local authorities in the face of apparently insuperable technical difficulties and inadequate financial resources compounded the difficulties.

Notes

1. *Census Reports*, 1871 and 1901.
2. Ibid. Most of tropical Asia, as one would expect, was too hot and wet for European comfort. While the size of the European community in Singapore increased significantly during the nineteenth century, it was largely male-dominated. Western women were wary of the heat and the propensity of being infected with tropical diseases. Few were ready to come with their husbands and to raise families. It was estimated that between 1820 and 1930, over 50 million Europeans crossed the oceans chiefly by steamship. The application of steam to oceanic travel made the passage overseas safer and cheaper than ever before. They were pushed from behind — the population of Europe was growing, but the supply of cultivable land was not. But there was also the matter of the pull, of the conviction held by these people that their lot would be better in the alien lands.
3. Arnold Pacey, *The Maze of Ingenuity: Ideas and idealism in the Development of Technology* (Cambridge, Massachusetts: The MIT Press, 1992), pp. 216–17.
4. Quoted in Ferguson, *Empire*, p. 168.
5. D. Cardwell, *The Fontana History of Technology* (London: Fontana Press, 1994), p. 290. The Crimean War also saw the use of the camera which "recorded accurately, without exaggerations of the poet and the imposed heroism of the artist, the scenes of fighting and the casualties after battle".
6. Quoted in Morris, *Pax Britannica*, p. 61.
7. Wright and Cartwright, eds., *Twentieth Century Impressions*, p. 39.
8. *Annual Report of the Straits Times*, 1859–60, p. 13.
9. *The Singapore Review and Monthly Magazine*, 1861–62, quoted in Walter Makepeace, Gilbert E. Brooke and Roland St. J. Braddell, eds., *One Hundred Years of Singapore* (London: John Murray, 1921; reprinted Singapore: Oxford University Press, 1991), Vol. 2, p. 149.

Introducing Technological Systems

10. Wright and Cartwright, eds., *Twentieth Century Impressions*, p. 45.
11. Quoted in Turnbull, *A History of Singapore*, p. 36.
12. "Review of the Progress of the Settlements since April 1867". Papers laid before the Legislative Council, 31 October 1873.
13. In 1873 John Pender amalgamated three telegraph companies, the British Indian Extension, The China Submarine and the British Australian, to form the Eastern Extension Australasia and China Telegraph Company (E. E. A. & C.). It was this company, combined with the Eastern, which formed the core of the telegraph empire established by John Pender. The first major project undertaken was the laying of a cable from Australia to New Zealand in 1876. Many local cable links followed, vastly improving the telegraph communications of the Southeast Asia area. The company remained in existence until 1974 when a joint company called Eastern Telecommunications Philippine Inc. was formed by Cable & Wireless and Philippine interests. This took over the assets of the E. E. A. & C.
14. Daniel Headrick, *The Invisible Weapon: Telecommunications and International Politics 1851–1945* (New York: Oxford University Press, 1991), p. 32.
15. Paul M. Kennedy, "Imperial Cable Communications and Strategy 1879–1914", *English Historical Review* 86 (1971): 730.
16. Ibid., p. 731.
17. Ibid., p. 735.
18. Robert V. Kubicek, *The Administration of Imperialism: Joseph Chamberlain at the Colonial Office* (Durham, N.C.: Duke University Press, 1969), pp. 30–33.
19. F. Swettenham, *British Malaya* (London, 1948, reprinted London, 1955), p. 175.
20. "To Singapore in a Second!", *British Malaya*, February 1929, p. 259. The eleven stations were Penang, Colombo, Seychelles, Aden, Port Sudan, Suez, Alexandria, Malta, Gilbrator, Carcavellos, and Porthcurnow.
21. *Report of the Telegraph and Telephone Communications Committee*, 31 August 1931, p. C258.
22. Telecommunication Authority of Singapore, *A Hundred Years of Dedicated Telephone Service in Singapore* (Singapore: International Press, 1979), pp. 5–9.
23. *Straits Settlements Progress Report 1930*, No. 73, 29 September 1930, p. C294.
24. *Report of the Telegraph and Telephone Communications Committee*, 31 August 1931, p. C251.
25. Buckley, *An Anecdotal History of Old Times in Singapore*, pp. 63, 363 and 398.
26. *Annual Report of the Straits Settlements, 1894*, p. 111.

27. Ibid. As classified by Britain's Ministry of Transport, Class One roads were the main traffic arteries and Class Two roads were routes of lesser importance.
28. *Annual Report of the Straits Settlements, 1897*, p. 314.
29. Ibid., p. 382.
30. G.E. Bogaars, "The Tanjong Pagar Dock Company, 1864–1905", *Memoirs of the Raffles Museum* 3 (1956): 178.
31. Daniel Cardwell regards the vogue of the bicycle as "a remarkable social and technological event of the last third of the nineteenth century" in Britain. The bicycle "encouraged women as well as men to take to the road [and] the social and political consequences of the bicycle, particularly as regards the progressive emancipation of women, have never been fully assessed". The bicycle boom led to improvement in road-building, the provision of wayside facilities for cyclists and the decline in the price of bicycles. See Cardwell, *History of Technology*, p. 367.
32. Raleigh Street, Nottingham, was the site of a small workshop which in 1886 started producing diamond-frame safety bicycles at the rate of three a week. Frank Bowden, a successful lawyer and convert to cycling, bought the firm in 1887 and in December 1888 founded The Raleigh Cycle Company as a limited liability private company. It grew rapidly and within a few years was a large public company capitalized at £100,000 (equivalent to about £5m today).
33. *Annual Report of the Straits Settlements*, 1893, p. 21.
34. For a study on railway development in Malaya, see Amarjit Kaur, *Bridge and Barrier: Transport and Communications in Colonial Malaya, 1870–1957* (Singapore: Oxford University Press, 1985).
35. "jinricksha" or "jinrickisha" is a Japanese term derived from "jin" meaning "man", "riki" meaning "strength, power" and "sha" meaning "carriage". The rickshaw first arrived in Shanghai from Japan in 1874 and would eventually become an iconic mode of transportation in that city and Beijing.
36. F.J. Hallifax, "Municipal Government", in *One Hundred Years of Singapore*, edited by Makepeace et al., Vol. 1, pp. 331–32.
37. The authoritative work on the rickshaw coolies in Singapore is James F. Warren, *Rickshaw Coolie: A People's History of Singapore 1880–1940* (Singapore: Singapore University Press, 2003).
38. Edwin A. Brown, *Indiscreet Memories: 1901 Singapore through the Eyes of a Colonial Englishman* (Singapore: Monsoon Books, 2007), p. 72.
39. *Proposed Jinrickisha Ordinance*, No. 21, 8 August 1899. Paper laid before the Legislative Council of the Straits Settlements.
40. Brown, *Indiscreet Memories*, pp. 73–75.
41. Quoted in Warren, *Rickshaw Coolie*, p. 74.

Introducing Technological Systems **117**

42. Historians of technology generally find it difficult to pinpoint the inventor of the automobile. The automobile is not so much a technology but as a technological system, and not a fixed technological system but one that has changed over time. Many nineteenth century inventors took up the challenge of creating a self-propelled road-running contraption seen as a superior substitute to horse drawn wagons and carriages. Henry Ford is generally credited with developing the first high-quality mass-produced car intended for the mass market. This is perhaps the reason why many regarded him as the inventor of the automobile. What Ford pioneered was a new method of assembly. See Eugene H. Weiss, *Chrysler, Ford, Durant, and Sloan: Founding Giants of the American Automotive Industry* (Jefferson, NC.: McFarland, 2003).
43. "Railways, Roads and Shipping Facilities in the Malay Peninsula", in *British Empire Exhibition Pamphlets*, Series Nos. 1–9, 1928.
44. *Annual Report of the Straits Settlements, 1920*, p. 47.
45. Eric Macfadyen, "Other Times, Other Ways", *British Malaya*, June 1926, p. 45.
46. "Engineering in Malaya", in *British Malaya*, January 1941, pp. 148–49.
47. *Malay Weekly Mail*, quoted in *British Malaya*, March 1939, p. 275.
48. J.R.H. Sidney, *Malay Land* (London: Cecil Palmer, 1926), p. 28.
49. A.F. Staples, "Malaya Revisited" in *British Malaya*, October 1935, p. 138.
50. Buckley, *Old Times in Singapore*, p. 710.
51. C.E. Wurtzburg, "Singapore: Past and Present", *British Malaya*, December 1940, p. 124.
52. Paper laid before the Legislative Council of the Straits Settlements, 2 November 1893, No. 29.
53. "Report of Burstall and Monkhouse", *Administrative Reports of the Singapore Municipality*, 1899.
54. "Corbet Woodall's report on the Electric Lighting in Singapore and the Proposed Acquisition of the works of the Singapore Gas Company", *Administrative Reports of the Singapore Municipality*, 1900.
55. The development of central power was the work of the least two decades of the nineteenth century. It was a tremendous technological achievement, made possible only by almost a century of large and small theoretical advances and practical innovations. Key development include: Volta's chemical battery in 1800, Oersted's discovery of electromagnetism in 1820, Ohm's law of electric circuit, experiments of Faraday and others, the invention of the electromagnetic generator in 1866 and the development of alternators and transformers for the production and conversion of high-voltage alternating current in the 1880s.

56. Bill Bryson, *At Home: A Short History of Private Life* (London: Doubleday, 2010), p. 124.
57. *Report of the Committee appointed by His Excellency the Governor of the Straits Settlements to Enquire into the Report on the Present Traffic Conditions in the Town of Singapore*, 29 August 1938, C.O. 76.
58. *Singapore Free Press*, 21 October 1936.
59. Ibid. The word "tidapathy" is likely to a combination of the Malay word "tidak apa" which means "cannot be bothered" with the English word "apathy".
60. The *Straits Times*, 18 August 1936.
61. Quoted by Mary Turnbull, *The Straits Settlements, 1826–67: Indian Presidency to Crown Colony* (London: Athlone Press, 1972), p. 40.
62. *Annual Report of the Straits Settlements, 1890*, pp. 5–6.
63. Ibid., p. 6.
64. Ibid., p. 6.
65. Ibid., p. 6.
66. Buckley, *Old Times in Singapore*, p. 420.
67. In ancient times, the Assyrians and Babylonians used clay to bind stones into a solid mass, and the Egyptians advanced a step further when they discovered the use of lime and gypsum mortar as a binding agent for building such structures as the pyramids. The Greeks made further improvements and finally the Romans developed a cement mortar that produced structures of remarkable durability. The secret of Roman success in making cement was traced to the mixing of slaked lime with pozzolana, a volcanic ash from Mount Vesuvius. This process produced a cement capable of hardening under water. During the Middle Ages this art was lost and it was not until the scientific spirit of inquiry revived that the world rediscovered the secret of hydraulic cement — cement that will harden under water. In 1824, Joseph Aspdin, a bricklayer and mason in Leeds, England, took out a patent on a hydraulic cement that he called Portland cement because its colour resembled the high-quality building stones quarried on the Isle of Portland off the British coast. Aspdin's method involved the careful proportioning of limestone and clay, pulverizing them, and burning the mixture into clinker, which was then ground into finished cement. Portland cement today, as in Aspdin's day, is a predetermined and carefully proportioned chemical combination of calcium, silicon, iron, and aluminum.
68. *Straits Times*, 17 April 1905.
69. Norman Edwards, *The Singapore House and Residential Life, 1819–1939* (Singapore: Oxford University Press, 1991), p. 214.
70. Ibid., p. 215.
71. Thomas R. Metcalfe. "Imperial Towns and Cities", in *The Cambridge*

Illustrated History of the British Empire, edited by Peter Marshall (Cambridge: Cambridge University Press, 1996), p. 237.
72. Quoted in Allen, *Plain Tales*, p. 521.
73. See Edwin Lee, *Historic Buildings of Singapore* (Singapore: Preservation of Monuments Board, 1990).

5
Sanitation and Public Health

Spreading industrialization in the West created new kinds of environmental issues. Early British and American factories spewed out smoke and were often condemned, as in Charles Dickens' works, for their ugliness. Water quality was also a serious issue. The sheer growth of cities, which often had unprocessed sewage running into local rivers, and the growth of the chemical industry with its cost-saving impulse to dump industrial waste products into the same rivers, produced noticeable health hazards by the end of the nineteenth century. Active concern for public health in Britain dates from 1840s, though it was not until the 1870s that the application of science and technology to public health problems began to be significant. It was generally thought that the solution to public health problems depended on improved sanitation provided by large-scale civil engineering projects. Later, science and medicine began to exert an ever-increasing role, especially from the 1880s with the discoveries of Louis Pasteur and Robert Koch and the germ theory assumed importance in the fight against communicable diseases.[1] As the nineteenth century progressed, technology and industrialization ushered in modern hygiene. Clean water, water closets, sewers, soap, together with better nutrition and improved housing, vastly reduced the incidence of infectious disease. Factories in the industrialized West mass produced cheap cotton underwear that was easy to clean.

While nineteenth century Singapore did not have industrial smokestacks to pollute the air and water, the rapidly growing colony and its population had to face up with environmental hazards that came with urbanization. The colony was shaping to be one of the many, to use the term favoured by urban historians, "walking cities" of the industrialized world. Since most inhabitants could not afford either the cost or the space required to keep a horse and carriage, they had to be able to walk to work or to work in their own homes. Businesses too had to be within walking distance of each other and this implied that as the colony grew, it became congested. More and more people had both to live and to work within the same relatively limited space. For most of the nineteenth century, building laws and by-laws to regulate town planning hardly existed. Some controls to monitor the construction of new houses and the rebuilding of old ones were introduced in 1896 but, generally, the government failed to prevent the haphazard growth of the city.[2] The passing of the Municipal Ordinance in 1887 did not result in massive land development because owners had only to pay a nominal tax on their undeveloped plots and, thus, were in no hurry to start construction.[3] Meanwhile, houses were not being built fast enough to accommodate the increasing population. The various by-laws and regulations introduced by the Municipality failed to impress upon the minds of the general public the importance of the science of sanitation. Overcrowding in the municipal area became inevitable and, consequently, morbidity and mortality from infectious diseases (in particular, tuberculosis and diarrhoeal diseases), remained high and continued to be so way into the 1950s.[4] Multi-ethnic colonial cities like Singapore suffered from ethnic stratification. At one end of the scale were the European officials and businessmen; at the other end were the thousands of unskilled new immigrants. Within each ethnic group there was further stratification, as in the case of the rich Chinese or Indian merchants and their poor and uneducated counterparts. Colonialism enhanced the prevalence of class distinction and structure, ghettos and ethnic neighbourhoods.

SANITATION AND SEWERAGE IMPROVEMENTS

Mid-nineteenth century England's "sanitary idea" made popular the notion that the physical living environment exercised a significant influence over the well-being of the individual and that the health depended upon sanitation. The late nineteenth century marked the rise of sanitary engineering as a new social profession in the industrialized West. Sanitary engineers possessed a comprehensive knowledge of the urban ecosystem and developed modern refuse collection and disposal methods and technologies.⁵ In the wake of the bacteriology revolution of the late nineteenth century, many health professionals shifted their focus from environmental sanitation and the miasmatic or filth theory to the germ theory of disease transmittal. While the demise of the filth theory led to a rethinking of the place of environmental sanitation within the municipal health departments, health workers generally recognized that refuse and waste disposal and poor sanitation practices were a health problem. The engineer was now able to find his place in supporting the rapid expansion of industrialized cities and offered their expertise in solving urban problems to municipal authorities. By the early twentieth century, sanitary engineers superseded health officers as leaders of refuse reforms in Britain and the United States. The need for safe water supplies, adequate sewerage, well-ventilated housing, and efficient refuse collection and disposal required the engineers' technical expertise and the public health officers' knowledge of sanitation. In Britain, during the 1870s, a hybrid profession — sanitary engineering — emerged to try to meet the environment challenge of the burgeoning industrial cities.

Municipal planning and public works in colonial cities like Singapore, Hong Kong and Calcutta were motivated by the growing interest in sanitation and the pressing need to accommodate a disparate population which, particularly in the case of Singapore, was stratified by various ethnic groups, living in demarcated ethnic neighbourhoods. Imported urban technologies and Western ideas of environmental aesthetics, however, tend to exacerbate this segregation. For the European community and those locals who had the means to emulate the lifestyles of the Europeans, a growing concern for hygiene led to the demand for clean water supply and modern sanitation system. However, during much of the nineteenth century, as recorded by a contemporary observer, the "public health of the town seems to have been left to look after itself for many years [and] epidemics of disease fortunately never became serious".⁶ The colonial administration was often criticized by the local newspapers for their tardiness in improving the urban environment. The passing of the Municipal Act of 1856 created a regular Municipal Council but it was only with the 1887 Municipal Ordinance, and the subsequent setting up of the Health Department in 1889, that proper planning and implementation of public works were carried out. In 1884, Frederick Weld, the Governor of the Straits Settlements, highlighted the effort made by his government:

We are, however, energetically urging on the work of sanitary improvement. A fetid and useless canal has been filled up, and a broad road substituted. The river, crowded nightly with dwelling boats, is being dredged and improved. Great foreshore reclamations are being made, and substantial public buildings have been and are yet being erected. New bridges are being built, trees planted, and drainage attended to by the Municipality. Private enterprise is also active, and one who has not visited Singapore for the last three years would hardly recognise it.[7]

The arrival of municipal engineers, beginning with J.W. Reeve in 1858 and followed by W.T. Carrington, Howard Newton, T.C. Cargill, James MacRitchie, and S. Tomlinson and R. Peirce did much to give the rapidly growing city a good urban environment, enhanced with the usage of imported sanitary knowledge and technologies. However, in the crowded Chinatown, thousands of Chinese immigrants packed into the tenements lining the narrow streets continued to live in unsanitary conditions. Despite an increase in the expenditure accounts of the Municipality during the years 1856 to 1888 which showed a ninefold increase from $62,799.90 to $539,097.55, the reality is that municipal authorities in the peripheral colonies of the Empire operated within a cost-conscious framework.[8] Adopting a *laissez-faire* approach to works, the port-city had to endure slow improvements in three main areas — conservancy, sewage and water supply — during the nineteenth century.

In 1827, Singapore had about 1.6 kilometres of drains completed in the town, to be "shared" by a population of 13,732. Street cleansing and the removal of domestic and street waste — household wastes, ashes, horse droppings, street sweepings, garbage, rubbish and even dead animals — were done in open carts. The street refuse was tipped, sometimes discriminately, onto sites near the river and onto vacant waste land. There was no further significant development in refuse collection and disposal until the late 1880s. The colony had to wait for industrialized Britain to develop and transfer knowledge and technologies that dealt with environment health. The "workshop of the world" led the way in disposal technology because many of its major cities, such as London and Manchester, had limited open space for dumping, and constraints on sea dumping due to English maritime interests.[9] Within these crowded cities, sanitation engineers advocated centralized systems of refuse disposal and the incinerator became an effective disposal technology. Singapore had its first incinerators built at Jalan Besar in 1889. The *Straits Times* reported in December of that year:

> The system of burning town refuse instead of burying it out of sight, now on its trial here, has successfully made way into public favour of late years in Europe, America and Australia. In this city, waste products and street rubbish have been largely availed of for reclamation purposes. The sanitary disadvantages of this method became too obvious to be disregarded any longer, and difficulties in the way of otherwise disposing of matter in the wrong place led to the suggestion to try incinerators, which, at the time, had answered in Calcutta.... Now that cineration has gained a footing here, it is hoped that the success attending it elsewhere in the world will declare itself as unmistakably in this city. The march of sanitary science demonstrates the danger of allowing refuse and rubbish to fester and decompose in and about towns.[10]

The incinerators at Jalan Besar were later supplemented, in 1910s, by additional ones built at Tanjong Pagar and Alexandra Road. These were the Heenan and Froude Incinerators developed in Worcester, England. By the turn of the century, all the refuse of the town, amounting to about 600 car-loads a day and cleared by about 2,000 coolies employed by the Municipality, was "scientifically destroyed".[11] Nevertheless, the rivers of Singapore continued to be indispensable dumping sites.

Until the implementation of a modern sewerage system, human waste in such crowded living environments was deposited in privy vaults in cellars or close to the house. The vaults were rarely watertight and produced noxious smells, as evident from a letter by "Daily Sufferer" in the *Straits Times*, dated 3 January 1907:

> I and my family are residents of Tanjong Katong, and I would ask your kind assistance in bringing to the notice of the Authorities the insufferable nuisance to which we are subjected every morning. I refer to the carts containing night soil ... These carts, although seemingly closed in, are by no means air-tight and the abominable stench emitted by them when passing, which poisons the air for long after they have gone by, make one sick....[12]

The human wastes were usually emptied at night, and thus the term "night soil" became a euphemism for human waste. The disposal of night soil using sewers "were said to be unsuitable for a town in the tropics with an ignorant population".[13] But there was a more scientific explanation to it, as given by MacRitchie, the municipal engineer (appointed in 1883), in an 1890 report:

> An ordinary system of water-borne drainage would, on account of the natural configuration of the Town and suburbs, the soft nature of the underlying soil, and the heavy rainfall, be attended with enormous cost, both in constructing the drains, and in maintaining powerful pumping engines to raise the sewage to a level to let it gravitate to the sea or to collecting tanks for treatment, and we should undoubtedly introduce that apparently necessary and attending bugbear of the system — sewage gas.[14]

Based on a study done in the English suburbs of Birmingham and Manchester, MacRitchie recommended the use of the Shone hydro-pneumatic system of excavation of sewerage by automatic ejectors worked by compressed air. In his second report produced in 1893, MacRitchie recommended a system of surface drainage for swill and storm waters and a two-pail system (the replacement of a clean pail each time a pail of night soil was collected) for excreta collection for eventual disposal at a sewage and poudrette plant set up in the same year. However, the municipal authorities clung tenaciously to its cost-conscious mentality, dragged its feet and no concrete actions took place until 1913 when a sewerage system was introduced. Till then, night soil was collected on a private basis and deposited at market gardens.

The provision of a safe water supply is essential to any satisfactory system of public health control. In the early days, the main sources of water were from wells. As decades passed and the population increased, wells continued to provide the water. Two technical advances made delivery of water more practicable (as happened in several key cities in Europe in early nineteenth century), namely, the application of steam power to water pumping and the use of cast-iron pipes which provided a durable, cost effective and technically manageable way to improve the distribution of water supply to individual structures. In 1852 John Turnbull Thomson proposed a waterworks scheme costing 28,000 pounds. However, the amount was considered too hefty to be defrayed by the revenue of the Straits Settlements.[15] While the hope of proper fresh water supply was left "tantalisingly dandled before the eyes of the native community", priorities were given to supplying fresh water to trading vessels at the port.[16] The same story of procrastination and differing priorities could be told of the provision of modern water supply and sanitary system. As reported in the *Daily Times*:

> The story of these Waterworks is anything but creditable to all concerned — Governors and Engineers — who have had to do with them. They must bear a share of the blame and discredit, from Governors Blundell and Cavenagh, with Col. Collyer and Captain Mayne under the Indian regime, to Sir Harry Ord and Major McNair under the Colonial office. It is the story of a continuous series of blunders and miscalculations…[17]

Eventually, between 1878 and 1918, several pieces of land were turned into compounding reservoirs and filters, service reservoirs and pumping stations. Outdated pumping stations with brick and wooden conduits were replaced by iron pipes when Singapore waterworks were handed over to the Municipality in 1876. Under the guidance of MacRitchie, new pumps, boilers and filters were installed by 1894 which increased the pumping capacity to 4,500,000 gallons of water per day. With the opening of the reservoir, many of the wells (about 4,000 by 1900) in Singapore were ordered to be closed. But there were complaints in the *Singapore Free Press* that the "reservoir was only a closed swamp, the water was sometimes greenish, gluey and soapy … [and the filters] were built in the worst place that could be found, near the Cemetery".[18]

The appointment of R. Peirce as the municipal engineer in 1900 was instrumental in effecting the "extension and improvement of the water supply and the installation of a water-carriage system for sewage for the city".[19] He reconstructed the majority of the town drains, enlarged the Thomson Road Reservoir (now MacRitchie Reservoir), and completed the Pearl's Hill Service Reservoir in 1907 (renamed as the Peirce Reservoir in 1911). The new century heralded in a period during which the Municipality played a proactive role in improving public sanitation and health.[20] Indeed, the municipal engineer became a central figure in the colonial administration's effort to manage, if not to protect, the environment. In Britain and America, the descriptive title "sanitary engineer" came to represent the nature of municipal engineering in the early decades of the twentieth century. He was recognized to have a relatively broad knowledge of the urban ecosystem, a working knowledge of engineering and public health theory and practices. In colonial Singapore, improvements to the living environment was given an impetus with the exhaustive report on the sanitary and health conditions of the town written in 1906 by W.J.

Simpson. In the report, Simpson stressed the need for the development of an efficient and comprehensive network of sewers to improve the well-being of the people. As recommended earlier by MacRitchie in 1890, Simpson also advocated the use of the Shone system of evacuation. Economic rationality influenced the municipal authorities and instead, gravitational sewers were laid, the contents of which were pumped to a distance and there disposed of in accordance with the latest scientific methods. As the colony celebrated the arrival of the new century, the system of water supply to the rapidly growing population began to improve tremendously. By the early decades of the twentieth century, water closets became more common in the urban residential areas.[21] After having been connected to the city-wide sewage system, the water closet contributed to a better urban environment. In the early years of its introduction, the water closet was "practicable in European houses and the larger institutions [but] they would be unworkable in the native quarters".[22] But, like all forms of scientific and technological hubris, the solution of one set of problems inevitably created other problems. The diffusion of the water closet to homes in rural areas led to an enormous increase in the use of water, which could not be met by the wells. Connections to the central waterworks had to be made. It also meant that the practice of using cesspools of human excrements by farmers as manures for their fields was slowly discontinued.

The sewage disposal problem raised the critical issue in sanitation science — the advantages of water filtration versus sewage treatment. Sanitary engineers argued for dilution and support for water filtration. Viewing sewage as a health problem, sanitarians and public health officials embraced treatment as the best safeguard against the spread of waterborne disease. In Britain, the scientific community began to apply principles of biology to water purification and sewage treatment. Among the various biological processes (such as the trickling filters system which was widely advocated in the early twentieth century) the development of the sedimentation septic tank provided the greatest versatility in the first step towards treating concentrated sewage. Its major value was the economic removal of sludge. Technological refinements were done and largely ended before 1914 with the development of a new tank designed by Dr Karl Imhoff of Germany. At the same time, a variety of processes were devised to aid in the treatment process, the most promising of which was the activated sludge process developed in the 1920s by Dr Gilbert Fowler of Manchester. By the 1930s, imported sanitation technologies had enabled Singapore to become "the cleanest city in the Orient".[23] Stereophagus Pumps driven by electric motors were installed in pumping stations, and Imhoff septic tanks, utilizing the activated sludge process, were used to purify effluents before they were discharged into the Singapore River.

But rapid urbanization came with a price. Tolerance of nuisances and fear of epidemic diseases played major roles in determining the sanitary quality of the Singapore community. A contributor to the *Straits Times* in 1889 wrote with a tinge of humour on his visitors to the island complaining vehemently of the stench and obnoxious smells of the environment as he accompanied them in a motor carriage, passing by "piggeries" and "duckeries", experiencing the "sour poisonous odours which floated across from sago factories" and the "deadly smell" of durian skins, Chinese tobacco, night soil and fish manure.[24] Decades later, the *Singapore Free Press* reported on the "smoke nuisance" in January 1928:

> From early morning till evening, and often through the night, those who have business offices along the seafront are smothered with a filthy layer of smuts blown in, when the wind so prevails, from the ships at anchor inside the breakwater ... coupled with the appalling stench from rubber factory operations, makes office-work on the seafront miserable. Nor is it only the shipping that the smoke nuisance emanates. It is a common place in town and the Municipality, through their incinerators and their pumping plants, etc., are as bad offenders as any private companies, while day by day, and at frequent intervals, the Traction Company makes a whole neighbourhood filthy through the huge clouds of fat-smoke belching from its chimney.[25]

Rapid urban growth of the Singapore city increased pressure on the municipal officials for effective environmental sanitation and threatened not only to further degrade the air, water and land but also to promote rounds of epidemic disease. One local newspaper reported that in March 1934, the quantity of refuse collected in 21,535 bins from about 37,000 premises, populated by 477,000 within the Municipal Limits, averaged 200 tons per day.[26] While the mechanization and motorization of street-cleaning equipment became popular by this time in the Western cities, mechanical sweepers were not commonly used in the Singapore roads and side streets. Although the Cleansing Department had a haulage fleet of 53 electric and motor vehicles, the high cost of these imported machinery and availability of cheap Indian coolies labour motivated the local press to conclude that "hand-sweeping is definitely cheaper and more practicable".[27]

Interestingly, the tasks of these road-sweepers were made more challenging during the durian and mangosteen seasons![28]

In the West, as mentioned above, faith in environmental sanitation as the primary weapon against disease lost support, especially in the medical community by the end of the nineteenth century. While the miasmatic theory failed to uncover the root cause of disease, it placed great emphasis on the need for environmental sanitation to combat it. The emergence of bacteriology, from the 1880s, brought changes to the way the world viewed health and disease.

TROPICAL MEDICINE AND IMPERIAL IDEAS

In the late nineteenth century, partly as a response to the perceived inability of Europeans to acclimatize the tropical regions of the British Empire and partly as a response to the increased contact between Westerners and indigenes, tropical medicine became institutionalized as a distinct branch of medicine in its own right.[29] Its rise was also magnified by the increasing economic and imperial importance of the colonial tropics for imperial powers like Britain and the United States. The tropics were seen as dangerous fever nests and uninhabitable by Europeans for any prolonged period. There was a fear of regeneration, that germs from the colonial world could be transported into the metropole.[30] Research and developments in tropical medicine were specifically directed towards selected diseases that threatened Europeans in tropical climates — and not including diseases that endangered the indigenous people. These efforts were carried out in the metropolis. Hence, although by 1900 medical research in Britain provided significant breakthroughs in combating tropical diseases, continual research within the colonies was never encouraged and, if initiated at all, was not all keenly received by local administrators.

Colonial Secretary Joseph Chamberlain was often credited for his effort in initiating the setting up of the London and Liverpool Schools of Tropical Medicine, an initiative which, in his view, was of great importance because it affected the "administration and wellbeing of the tropical colonies" and that he would "encourage by every means in his power, scientific enquiry into the causes of tropical diseases".[31] The London School of Tropical Medicine was opened in 1899 at the Albert Dock Hospital, one of the establishments of the Corporation of the Seamen's Hospital Society. It was sited in London precisely because London had ports which received merchant vessels arriving from countries in the tropics where diseases peculiar to the region were prevalent. The foundation of the organized study and treatment of tropical diseases was largely attributed to the ingenuity of Sir Patrick Manson, the "Father of Tropical Medicine" and Ronald Ross. But research and practical health measures were Eurocentric in nature; colonies covered all costs in order to provide a more conducive environment for Europeans to live in India.[32] Chamberlain's main concern was the "white man" living in the tropical colonies of the British Empire. As seen in his letter to the Government of the Straits Settlements, he reiterated the "important question of reducing the abnormal rate of mortality and sickness among European residents in tropical climates [and] from the same source ... the health of the masses will follow".[33] In her study of tropical health in colonial Malaya and Singapore, Lenore Manderson concludes that "the introduction of biomedical institutions, practitioners, treatments and cures were part of the colonising project that, by introducing Western cultural values to the empire, insisted on the moral authority of the imperialising power".[34] In many of the rubber plantation owned by British entrepreneurs and run by British plantation managers, estate coolies suffered from poor living conditions and poor nutrition which led observers like Manderson to identify these plantations as enclaves where workers were systematically exploited to benefit British businesses and industries. In these estates, "[o]wners and managers were reported to be absolutely indifferent to the welfare of their labourers, indentured or otherwise ... they were conservative and reluctant to invest in sanitation or to employ dressers to meet the most immediate needs of those injured or sick, insisting that the government was responsible for environmental control".[35] Under such exploitative environment, resistance to infections was very low.

While the *Straits Times* claimed in 1861 that the European community in Singapore was "the healthiest community in the East",[36] and John Cameron, writing in 1865, claimed that the climate of Singapore "has been established beyond all doubt to be kinder and more genial to the European constitution than any other area in the east",[37] the reality was that many European womenfolk suffered. Dr Robert Little, who came to Singapore in 1840 as a private practitioner and later became the Honorary Secretary of the Committee of Management for the Tan Tock Seng Hospital, wrote extensively on developments in the island and commented that the women found the climate trying and the heavy clothing and lack of physical activity further contributed to their plight. It was even asserted that lactation was difficult for European women living in the tropical climate of Malaya and very few women were able to nurse their babies for more than a week.[38] Western medical experts blamed the inhospitable tropical climate, as in Malaya and Singapore, for producing harmful effects on Europeans. Alfred Crosby explains the perils of living in the tropics:

> Most of tropical Asia, as one would expect, was too hot and wet for the European tastes, but most important than its propensity for making invaders sweat was the teeming presence of minute enemies. The Asians and their plants and animals had existed in and around thousands of villages and cities for thousands of years, and along with them had evolved many species of germs, worms, insects, rusts, molds, and what have you attuned on humanity and its servant organisms.[39]

Within the crowded city limits, the unhealthy living environment among the Asians was exacerbated by their unsanitary customs and habits. Food adulteration was a critical issue. To the Westerners, the concept of food hygiene in "eastern nations, it exists, if it does at all, only in theory, and it is difficult to convey to such minds that cleanliness is one of the positive commands arising from the constituted order of things, and that the absence of the practical knowledge of sanitation is one of the principal causes of dreadful diseases common among them…"[40] The "primitive conditions" of bakeries and dirty food-stalls were singled out by the health officers.[41] Although living standards continued to improve as the port-city entered into the twentieth century, food sanitation continued to be problematic, as highlighted in the *Singapore Free Press* in September 1923: "the people of these islands are too much inclined to assume that, as there are Government inspectors for milk, meat and other foods, they have no need to worry [and that] their food has been effectively protected from dirt and everything harmful — that is, their food has been effectively sanitated".[42] As late as the 1920s and 1930s, the unhealthy living environment was starkly seen in the crowded rickshaw and port coolie quarters where as many as 30,000 men lived in these lodgings most of which had neither running water nor toilets. Here, in the words of James Warren, "sickness announced itself every day with the rattling cough of the consumptive or the figure of the coolie squatting with diarrhoea over a filthy latrine; while many emerged each morning in wretch condition with a bout of fever".[43] Life was decidedly hard for the rickshaw coolie.

The first hospitals in Singapore were built as early as 1821 to serve the needs of the British officers and their troops.[44] For the Chinese paupers, the foundation stone of the Tan Tock Seng Hospital (built through the generous donation from the man himself) was erected on 25 July 1844 and started its operation only in 1849 due to the lack of funds to run the hospital. It was the first hospital to be built with non-Government funds. Access to medical and hospital services for the larger population was provided by the government on an *ad hoc* manner. The grim living conditions of the masses gave Singapore a mortality rate higher than in Hong Kong, India or Ceylon, ranging from 44 and 51 per thousand during the first few years of the twentieth century.[45] By a stroke of good luck, there were no plague epidemics but the population had to face up with endemic killing diseases, such as cholera, beri-beri, tuberculosis, malaria, enteric fever, dysentery and leprosy. The port-city was exposed to the ravages of cholera from the time of its founding in 1819 because of its intimate association with Bengal — regarded by the medical world as the most important epidemic centre of cholera in the world.[46] The disease is thought to have origins in India and emerged onto the world stage only in the nineteenth century. Interestingly, the speed of travelling as a result of faster steamship and the Suez Canal allowed the bacterium to survive and infected lands way beyond the Indian continent. Throughout the nineteenth century, cholera was endemic in Singapore and the Singapore press frequently exhorted the government to implement anti-cholera measures. In February 1862, the disease assumed epidemic proportion when forty-four Chinese died within a short period of four days.[47] The *Straits Times* hinted that the government was indifferent because no Europeans had been infected and that complacency would lead to more deaths amongst the natives: "We should be sorry to alarm the public and unless we wish by so doing to awaken it from an apathy which is dangerous in itself and in its consequences....Hitherto the maladies and afflictions … were not of a character that extended themselves to Europeans, yet it is impossible to say how soon some alarming epidemic might disclose itself, that would sweep away the native and the European alike…".[48] In most situations, epidemics forced the colonial administration to deal with public health more vigorously, but the absence of regularized preventive action (at least during the nineteenth century) had to do with limited knowledge about contagious diseases. Within the medical profession, many believed that the communicable disease was caused by contagion and that the germs could only be activated in conjunction with other elements, such as the state of the atmosphere or the condition of the soil. Measures to eradicate the disease included cleaning up the community, improving the sanitation system and quarantine of people.[49] For the more affluent Europeans, they could sail back to Europe to escape the scourge of disease. For the majority of the Chinese and other local inhabitants, death from disease was a sure sign that some evil spirits were at work and it was necessary to appease them. As reported in the *Straits Times*: "On 12th March 1862, the Commissioner of Police granted permission to the Chinese to fire off crackers for one hour each evening for three days. The Chinese believed that this drove away Cholera (the evil spirits causing it)".[50] The European community complained bitterly that "[t]he processions ... where men with steel knives pierced through their faces, are dragged about by a few of their countrymen beating gongs and waving flags, are sickening and repulsive to the eye and positively indecent, and should most certainly have been put a stop by the Police".[51] It was a typical perception — and fear — of the European community that the constant social interaction with the Asian masses would inevitably exposed them to the vagaries of infection and "filth" disease like cholera and tuberculosis. The Asians, particularly the Chinese who lived in sardine-packed tenements, were stereotyped as people who were totally complacent and ignorant of the danger of contamination, filth and dirt. Such a characterization of the Asians fostered the argument taken up by the Municipal Health Department that it would be waste of financial resources to provide modern sanitary facilities.[52]

At the start of the twentieth century, the standards of public health were not high, according to the Medical Report on the Colony by Dr Mugliston.[53] Malaria was regarded as the main killer, accounting for 1,410 out of 9,440 deaths in 1909. While the situation in Singapore was probably not severe due to the limited land available for rubber plantation, the prevalence of malaria in the Federated Malay States was extensive, particularly amongst the thousands of Indian coolies employed in the rubber industry.[54] Information on Western methods of anti-malaria control, ranging from the use of quinine to drainage and oiling, and articles relating to combating malaria, were disseminated to "estate hospitals" through various media — journals, magazines, newspapers, and, in the late 1930s, radio broadcasting. Despite the criticism against British plantation owners, several of the larger estates in the Straits Settlements and in British Malaya had set up their own "estate hospitals" which provided rudimentary and often inadequate facilities and qualified medical personnel. Liew argues that "an understated legacy of the planters in British Malaya [including the Straits Settlements] was their contributions towards the moulding of the study of malarialogy and the application of anti-malaria preventive public health work".[55] Plantation managers frequently engaged researchers and medical personnel to carry out preventive measures of diseases and infections. In this sense, "it was the plantations that expanded the application of knowledge in routine preventive health work [and] provided the colonial medical fraternity with sites to develop the discourse on tropical diseases".[56] In British Malaya, it was reported that in 1911 the death rate per thousand was 232, in 1923 it was 3 per thousand and, in 1938, the figure had fallen to 0.08 per thousand.[57] Administrators and missionaries operating in tropical colonies were also particularly concerned about leprosy. In 1885, the writings of Henry Wright, Archdeacon in the Church of England at Grantham propagated the "Imperial Danger" thesis that the dreaded medieval disease would once again be imported into England from India. In the West after the mid-nineteenth century, municipal health officials and sanitation engineers strove to separate the germs from the masses. In the tropical colonies, when the municipal authorities could not achieve this objective, they separated the infected people from the healthy ones. In the beginning of the twentieth century, leper confinement controls were introduced in Malaya, Singapore and colonies in Africa. Compulsory segregation of lepers in the Straits Settlements was introduced through the passing of the Lepers' Ordinance in July 1897. The neglect of lepers in the tropical colonies could be attributed to the attitudes of Western medical doctors who, according to one medical missionary, viewed leprosy "as an incurable infirmity, rather than a disease ... for fear of contracting the disease itself ... [few of them] were willing to undertake its treatment".[58] In the British-run leprosarium in Singapore, in the words of Sheldon Watts:

> medical doctors were also rare birds whose absence left patients to the attentions of lay staff. Asian warders used long sticks to prod lepers from place to place and gave them food in old tin cans slipped under the doors of their cages. In Singapore city, when rounding up lepers, attendants wore masks and gloves and liberally sprayed victims' houses and possessions with decontaminants.[59]

As in many places in the world, lepers were not released from the internment camp and there were no provision for discharge, except through death. The imperial response to leprosy "had become a distinctively punitive form of coercion in which segregation was very often for life ... and the institutionalization of the leper colony in the later nineteenth century, which developed into an extreme example of the boundary thinking than had come to dominate relations between colony and metropole".[60] Island leper colonies within the British Empire — such as those in the coastal regions of New Zealand and Australia, in the Caribbean and the Indian Ocean — were set up in colonies with a high proportion of European settlers. According to Loh, the stigma against leprosy was more serious in colonial Singapore "precisely because of its location on the great colonial trade and migration routes and because of its importance to the British Empire".[61]

Measures were also carried out to introduce systematically the science of midwifery, so as to increase the number of "trained" midwives — traditionally known as *bidans* for the Malays and *jiesheng fu* for the Chinese. They were trained to handle the needs at maternity hospital (with eight beds), built in 1888, and by 1907, had expanded to sixteen beds for "natives" and thirteen beds for European women.[62] Under the direction of Dr Middleton, the municipal health officer in Singapore, lectures on the importance of aseptic midwifery practice were given to midwives. In 1910, the Maternal and Child Health Services was set up as registration and treatment centre to provide maternity care for local women, especially those from poor families. The first European nurse was appointed in Singapore to gather information on early infant life and to educate women in child care. Female inspectors were also trained to make home visits to provide post-natal and infant care advice. However, modern science had to face up with staunch traditional practices and superstitious beliefs of daily living of the natives in Singapore, including midwifery practice. The persistent efforts of the health inspectors and the opening of health centres in the rural areas to provide public health education were significant factors which accounted for the gradual acceptance of Western midwifery methods and medication. In 1930, the Midwives' Ordinance was passed to promote midwifery practice, especially in the rural areas and, together with the spread of infant and maternal welfare clinics, the infant mortality rate declined from 183 per thousand in 1930 to about 165 per thousand by the end of the 1930s.[63]

On hindsight, and to be fair to the colonial officials in the municipal services, the changing economic, political and social conditions experienced by Singapore in the course of the nineteenth century produced difficult problems which were beyond what the officials could anticipate or do. Before the Transfer of 1867, the indifferent attitude of the East India Company towards the Straits Settlements was an inhibiting factor.[64] Singapore was also used as a dumping ground for convicts of Continental India and "with such a large body of convicts there is no adequate provision for the protection of the life and property of the inhabitants".[65] Such an approach could only but reinforced, firstly, the unwillingness of British officials to serve in the Settlements and, secondly, unnecessary heavy expenditures incurred by the Settlements were discouraged. As late as 1880s and even when Singapore was considered the "richest of the Crown Colonies under the British flag … a good many people at home (London) hardly even know where Singapore is".[66] Lured mainly by trade and profit, many left the shores of Britain and British India. The emergence of Singapore as the premier British port opened the floodgates to thousands of hopeful immigrants from China, India and the region, inflating the size of the population and creating tremendous pressures on the provision of sufficient social and health amenities. These developments were something which local officials could not easily anticipate.

The efforts of the municipal officials in providing public improvement schemes in Singapore could be appreciated a little more if seen in the light of such developments in Britain. Municipal engineering gained prominence in Britain only in the 1850s with the publication of Edwin Chadwick's massively influential "Report on the Sanitary Condition of the Labouring Population of Great Britain" in 1842.[67] Chadwick (a lawyer by training) believed that putrefying vegetables and human fecal matter released life-threatening fumes (miasmas) which led to a whole range of fever-like diseases. In tackling the unsanitary conditions of London, he recommended the development of an efficient sewerage system and that help must be sought from "the science of the Civil Engineers, not from the physician, who has done his work when he has pointed out the disease that resulted from the neglect of proper administrative measures".[68] But Dickens' London only enjoyed the benefits of "sanitary science" propagated by Chadwick and sanitary engineering when the Metropolitan Board of Works was established in London in 1855 and Sir Joseph Bazalgette was appointed chief engineer to the board. He carried out the construction of an extensive network of sewers in London (which is still at the basis of the city's sewage system in the twenty-first century), "diverting all of London's sewage from the Thames and channelling it eastward, through new systems of pipes to the north and south of the river, to treatment plants far outside the city in Essex and Kent" during the years 1855 to 1875.[69] The causal connection between sewage, water and cholera was only confirmed in 1883 when Robert Koch, a young German physician, discovered the *cholera bacillus*. What followed was a series of scientific investigations carried out at the turn of the century in Britain and the United States on the whole problem of water filtration, culminating in the less and more effective method of treating water with chlorine.[70] These ideas and the technologies associated with municipal engineering took sometime before they were transferred to the main colonial cities of the British Empire. In Hong Kong concrete improvements in sanitation and water-supply took place from the late 1890s; in Calcutta, the presence of a large and prosperous European community enabled the city to expand its sewerage and drainage system from the 1870s.[71] In Malaya, most of the health initiatives "were implemented soon after they were introduced in the centre, and reflected not government understanding of colonial needs, but rather contemporary changing ideas and practices in public health occurring within the United Kingdom".[72]

In March 1934, the *Singapore Free Press* declared that "Singapore is one of the healthiest cities in the world".[73] Although overcrowding of the city was a concern for the Municipal Council — in 1931, 447,741 (out of a population of 567,433), of which were 8,147 Europeans, lived within the municipal limits, giving a density of 14,000 persons per acre — living conditions continued to improve. Kenneth Black, Professor of Surgery, King Edward VII College of Medicine, stated in 1933: "Progress in the prevention and cure of tropical diseases has advanced enormously in the past twenty-five years, and, it may be argued, that Malaya is to be regarded 'healthy'".[74] Likewise, the colonial administrator, Richard Winstedt proudly wrote in 1935 that "British medicine practically extinguished small-pox, cholera, typhus, yews and diphtheria, banished malaria at least from towns and reduced tuberculosis and infant mortality".[75] While in the first instance the beneficiaries of the measures to implement better sanitation and healthcare were the British and other Europeans, and that the majority of the indigenous population lived in squalid conditions, it is fair to conclude that the colonial government was increasingly committed to improving the health of the masses. In the years after 1945, the major challenge for municipal officials, engineers and town planners was to adapt sanitary services to growth characterized by rapid suburbanization, on the one hand, and demand for such services in the rural communities, on the other.

Notes

1. See Sheldon Watts, *Disease and Medicine in World History* (New York and London: Routledge, 2003), Chapter 9.
2. Teo Siew Eng and Victor Savage, "Singapore Landscape: A Historical Overview of Housing Image", in *A History of Singapore*, edited by Ernest Chew and Edwin Lee (Singapore: Oxford University Press, 1991), pp. 312–38.
3. Ibid., p. 325.
4. For an examination of how the urban built environment of colonial Singapore was shaped between 1880 and 1930 by conflict and negotiation between the Municipal Authority of Singapore and the indigenous communities, see Yeoh S.A. Brenda, *Contesting Space: Power Relations and the Urban Built Environment in Colonial Singapore* (Singapore: Oxford University Press, 1996).
5. See Martin Melosi, *Garbage in the Cities: Refuse, Reform and the Environment* (Pittsburgh: University of Pittsburgh Press, 2005), especially Chapter 3.
6. Quoted in Makepeace et al., eds., *One Hundred Years of Singapore*, Vol. 1, p. 321.
7. Quoted in Paul Kratoska, *Honourable Intentions: Talks on the British Empire in South-East Asia Delivered at the Royal Colonial Institute, 1874–1928* (Singapore: Oxford University Press, 1983), p. 50.
8. Ibid., pp. 318–19.
9. Martin Melosi, *The Sanitary City: Environmental Services in Urban America from Colonial Times to the Present* (Pittsburgh: University of Pittsburgh Press, 2008), p. 124.
10. *Straits Times*, 31 December 1889.
11. Quoted in Makepeace et al., eds., *One Hundred Years of Singapore*, Vol. 1, p. 325.
12. *Straits Times*, 3 January 1907.
13. Quoted in Makepeace et al., eds., *One Hundred Years of Singapore*, Vol. 1, p. 325.
14. "Report on the Estimates for the Disposal of Night Soil", *Annual Report of Singapore Municipality*, 6 September 1890.
15. Ibid.

16. Ibid.
17. *Daily Times*, 1872, quoted in Moore and Moore, *The First 150 Years of Singapore*, p. 382.
18. *Singapore Free Press*, 12 October 1909.
19. Quoted in Makepeace et al., eds., *One Hundred Years of Singapore*, Vol. 1, p. 321.
20. See Kuldip Singh, "Municipal Sanitation in Singapore, 1887–1940", unpublished honours thesis, Department of History, National University of Singapore, 1989.
21. The water closet was first introduced in England in 1810 and the technological wonder offered city dwellers a more convenient and more sanitary method of disposing of human waste. It encouraged greater use of water and, when linked to cesspools, reduced the effectiveness of the cesspools. Due to the high water usage, cesspool waste did not percolate into the soil but overflowed the cesspools and found its way into the streets and city drainage systems.
22. *Eastern Daily Mail*, 23 September 1907. This was the opinion of the Health and Disposal of Sewage Committee in response to the Simpson Report.
23. *Singapore Free Press*, 22 March 1934.
24. *Straits Times Weekly Issue*, 15 August 1889.
25. *Singapore Free Press*, 10 January 1928.
26. *Singapore Free Press*, 22 March 1934.
27. Ibid.
28. Ibid.
29. Rod Edmond, *Leprosy and Empire: A Medical and Cultural History* (Cambridge: Cambridge University Press, 2006), Chapter 3.
30. Ibid.
31. It was estimated by Chamberlain that to set up the School of Tropical Medicine, 7,000 pounds sterling was needed. In his letter to the Government of the Straits Settlements, he requested this amount would be met by the British Possessions in West Africa, Gold Coast, Lagos, Niger Coast Protectorate, the Federated Malay States and the Straits Settlements (which had to contribute about 650 pounds sterling). In another letter to the Governor of the Straits Settlements, dated 28 February 1899, Chamberlain acknowledged the contribution of "$5,000 towards the cost of the contemplated School of Tropical Medicine and the Commission appointed to study the subject of Malarial Fever, and that the Federated Malay States have agreed to make an equal contribution". See "School of Tropical Medicine", Paper laid before the Legislative Council of the Straits Settlements, No. 12, 14 February 1899, "Contribution towards the Cost of the contemplated School of Tropical Medicine and Malarial

Fever Commission", Paper laid before the Legislative Council of the Straits Settlements, No. 19, 25 April 1899, and also "London's Hospital for tropical Diseases", *British Malaya*, September 1929, pp. 149–55.

32. Richard Kesner, *Economic Control and Colonial Development: Crown Colony Financial Management in the Age of Joseph Chamberlain* (Oxford: Clio Press, 1981), p. 153.
33. Quoted in Francis Fremantle, *A Traveller's Study of Health and Empire* (London: John Ouseley Ltd., 1911), p. 346.
34. Lenore Manderson, *Sickness and the State: Health and Illness in Colonial Malaya, 1870–1940* (Melbourne: Cambridge University Press, 1996), p. 231.
35. Ibid., pp. 136–37.
36. *Straits Times*, 17 August 1861.
37. John Cameron, *Our Tropical Possessions in Malayan India* (London: Smith Wilder, 1865; reprinted, Kuala Lumpur: Oxford University Press, 1965), p. 249.
38. A. Castellani, "The Adaptation of European Women and Children to Tropical Climates", *Proceedings of the Royal Society of Medicine* 24 (1931), pp. 95-98.
39. Alfred W. Crosby, *Ecological Imperialism: The Biological Expansion of Europe, 900–1900* (Cambridge: Cambridge University Press, 1986), p. 135. The theme of environment and biological science as an argument for the advance of imperialism is the focus of this study. Crosby argues that the success of Europeans' conquest of the native peoples in the temperate zones was due more to a broader set of environmental considerations than of military conquest. For a specific reference to the countries in Southeast Asia, see Victor Savage, *Western Impression of Nature and Landscape in Southeast Asia* (Singapore: Singapore University Press, 1984).
40. *The Daily Advertiser*, 17 July 1890.
41. Ibid.
42. *Singapore Free Press*, 3 September 1923.
43. Warren, *Rickshaw Coolie*, p. 259.
44. The most comprehensive coverage of Singapore's medical history from the 1820s to the 1870s was done by Lee Yong Kiat, *The Medical History of Early Singapore* (Tokyo: Southeast Asian Medical Information Center, 1978).
45. Turnbull, *History of Singapore*, p. 113.
46. Lee, *Medical History*, p. 239.
47. *Straits Times*, 1 March 1862.
48. Ibid.
49. An international conference of cholera experts was held in Constantinople in 1866 at which serving English doctors in India believed that the spread

of the disease could be prevented by restricting the movement of suspect cholera carriers. It was not till the late 1880s and with the persistent efforts of Florence Nightingale that the medical field eventually paid heed to her claim that the health of India's rural population was being undermined by unsanitary conditions. Nightingale was a firm supporter of the notion that cholera was not at all a contagious diseases and that quarantine procedures to prevent its spread were unnecessary.

50. *Straits Times*, 15 March 1862. For a short account of the link between religion, tradition and health, see David Clark, *Germs, Genes and Civilization: How Epidemics Shaped Who We Are Today* (New Jersey: Pearson Education, 2010), Chapter 8.
51. *Straits Times*, 5 April 1862.
52. See Yeoh, *Contesting Space*, Chapter 3.
53. *Straits Times*, 13 September 1902.
54. See Liew Kai Khiun, "Planters, Estate Health & Malaria in British Malaya 1900–1940", *Journal of the Malaysian Branch of the Royal Asiatic Society* 83, Part 1 (2010): 91–115.
55. Ibid., p. 83.
56. Ibid., pp. 110 and 112.
57. "Mosquito Day Luncheon", in *British Malaya*, June 1939, p. 43.
58. Watts, *Disease and Medicine*, p. 70.
59. Ibid.
60. Edmond, *Leprosy and Empire*, p. 142.
61. Loh Kah Seng, *Making and Unmaking the Asylum: Leprosy and Modernity in Singapore and Malaysia* (Petaling Jaya: Strategic Information and Research Development Centre, 2009), p. 18. Loh's case-study of leprosy and the institution of the leprosarium from a historical and contemporary perspective provides a useful understanding of the social impact of modern science and medicine on a society in the *longue duree* — from a British colonial regime in the beginning of the nineteenth century to an independent, modern Singapore of the twenty-first century.
62. See Lenore Manderson, "Blame, Responsibility and Remedial Action: Death, Disease and the Infant in Early Twentieth Century Malaya", in *Death and Disease in Southeast Asia*, edited by Norman Owen (Singapore: Oxford University Press, 1987), Chapter 12, for a succinct account of infant healthcare and midwifery during the 1900 to 1930s in the Straits Settlements.
63. *Straits Settlements Progress Report 1930*, No. 73, 29 September 1930, p. C289.
64. In 1867, control and administration of the Straits Settlements was transferred from the Colonial Office in India to the Crown Office in London.

65. *Petition to Parliament in 1858 on the Question of Transfer of the Straits Settlements to the Crown Office*, quoted in *The First 150 Years of Singapore*, edited by Moore and Moore, p. 313. Strangely, it was reported in the Legislative Council of the Straits Settlements that "it is noteworthy that when the time came for their departure a strong opinion was generally expressed in favour of their retention in the Colony". See *Straits Settlements Progress Report, 31 October 1873.*
66. Quoted in Kratoska, *Honourable Intentions*, p. 82.
67. W.H.G. Armytage, *A Social History of Engineering* (Faber and Faber, 1976), p. 140. The Report maintained that disease caused poverty — precise reverse of the position that later would be held by the German medical pioneers Rudolf Virchow and Robert Koch.
68. Ibid.
69. Michael Paterson, *Voices from Dickens' London* (Cincinnati: David and Charles, 2006), p. 31. See also J. Rawlinson, "Sanitary Engineering: Sanitation", in *A History of Technology*, edited by Charles Singer (Oxford: Oxford University Press, 1958), Chapter 4.
70. James Finch, *Engineering and Western Civilization* (New York: McGraw Hill, 1951) p. 238.
71. Headrick, *Tentacles of Progress*, pp. 150, 157.
72. Manderson, *Sickness and the State*, p. 11.
73. The *Singapore Free Press*, 22 March 1934.
74. Kenneth Black, "Health and Climate with special reference to Malaya", *British Malaya*, March 1933, p. 253.
75. Richard O. Winstedt. *A History of Malaya* (Kuala Lumpur and Singapore: Marican & Sons, 1982), p. 266.

6
Agriculture and Colonial Science

During the nineteenth century, Western imperialists recognized the tremendous wealth — and power — that could be accrued from a systematic implementation of economic botany in their colonies. They were competing to discover, categorize, propagate, and exploit the wealth of plants from Asia and the Americas that had suddenly become available to them. By the late seventeenth century, the undisputed leaders in the field of economic botany were the Dutch, who had pushed aside the Portuguese to become the dominant European power in the East. The Dutch wanted to understand new plants for two main reasons: to find cures for the tropical diseases that were afflicting their sailors, colonists and merchants; and to find new agricultural commodities beyond the known spices, from which new wealth could be generated. They set up botanical gardens at their colonial outposts at the Cape, at Malabar, Ceylon and Java, and in Brazil, all which exchanged specimens with botanical gardens in Amsterdam and Leyden. Soon, England and France joined in the race. In this new field of "economic botany", the pursuit of scientific knowledge went hand in hand with furthering the national interest. Botany was considered as "big science" during this period, an indicator of a country's might and sophistication. Indeed, some botanical gardens were even laid out in four quadrants, each representing Europe, Africa, Asia and the Americas. They were further subdivided into individual beds for particular plants. It was the dream of the imperial botanist to gather the world's plants in one place. In England, the Kew Gardens, set up in 1772, became the centre for botanical research in the British Empire. Around Kew revolved botanical stations set up in some colonies. These gardens were interlocked by a constant exchange of ideas, live plants and expertise. The objective of these colonial gardens was not, of course, the delectation of the inhabitants with beautiful flowers and trees. They were set up for the very practical purpose of experiments with new crops. Through its scientific study of plants transferred, the Kew Gardens and its satellite gardens in the colonies converted knowledge to economic gains and power which, in turn, contributed to making Britain as the number one industrial power in the world.[1] Robert Kyd, a British army officer and founder of the Calcutta Botanic Gardens in 1787, recorded the gardens were established "not for the purpose of collecting rare plants as things of curiosity or furnishing articles for the gratification of luxury, but for establishing a stock for disseminating such articles as may prove beneficial to the inhabitants, as well as the natives of Great Britain, and which ultimately may tend to the extension of the national commerce and riches".[2] This was made possible because of a tightly integrated system Kew developed with its satellite stations. The former was the control and nerve centre where plants with economic potential were carefully propagated in the Kew greenhouses and, with and other related scientific information of commercial importance, were channelled to colonial gardens. The latter carried out developmental work on the economic plants, educate local planters, distribute the plants and published findings in bulletins. This relationship was extremely important for the success of economic botany. In the tropical colonies, two types of tropical agriculture were experimented and exploited: peasant agriculture and plantations. The former involved coercing, either by force or promises of economic benefits, peasants to grow commercial crops for export. The Culture System of the Dutch East Indies was probably the outstanding example in Southeast Asia, though it cannot be denied that the roots of Holland's competitive advantage in the flower industry today were laid in a series of aggressive but scientifically rewarding agricultural research policy in Java. Plantation agriculture, on the other hand, required the compromising role of the government in introducing changes such as irrigation systems, better seeds, agricultural education, and marketing schemes in order to develop export crops.

Despite the dominance of commerce as the main economic lifeline in Singapore's economic growth, a modest start was made in the application of science to the production of tropical export crops during the early decades of the nineteenth century. However, it is argued here that the early exposure of the local community to developments in agricultural science and botany failed to create an indigenous scientific culture during the colonial period.

ESTABLISHING THE BOTANIC GARDENS

In Singapore, in 1822, Stamford Raffles gave forty-eight acres of land to Dr Nathaniel Wallich of the Botanical Gardens at Calcutta to establish a Botanical and Experimental Garden, for "the purpose of forming a depot for plants, from the circumjacent parts of the world, for which purpose Singapore seemed admirably adapted by its central situation and mild climate, by means of which plants indigenous to countries situated in high attitudes, might more easily be naturalized to a tropical climate, at the same time the experiment of cultivating spices was to be tried and if found to succeed might be carried on to any extent necessary."[3] Born at Copenhagen, Denmark, on 28 January 1786, Nathaniel Wallich had a distinguished career as a surgeon and botanist with the East India Company's Botanical Garden at Calcutta. He had a hand in charting the course of British history in the Far East. In 1835, Wallich examined some fresh seeds which were smuggled into Assam (in northern India) from China. He confirmed them to be *camellia sinensis* — the tea leave which became an instrument of British imperial control and its cultivation impacted on Britain's Industrial Revolution in the second half of the nineteenth century.[4] From the early eighteenth century tea began to be imported in sufficient quantities to create a mass market in Britain. Home consumption leapt from an average of under 800,000 lbs in the early 1740s to over 2.5 million lbs between 1746 and 1750.

After the signing of the Anglo-Dutch Treaty in 1824, Dutch supremacy in the Indonesian Archipelago was officially recognized and they now possessed a virtual monopoly of the valuable spice trade.[5] The experimentation and cultivation of spices by the British became all the more necessary. Moreover, Singapore "had a great extend of ground quite unoccupied and well adapted for the purpose".[6] Such optimism was reinforced by Wallich's own study of the southern bank of the Singapore River in which he pointed out the richness of the soil and the suitability of the climate for the experimental cultivation of indigenous plants of Singapore.[7] In November 1822, forty-eight acres of land on the slope of Fort Canning became the site of the Singapore Botanic Gardens. In January 1823, Wallich left Singapore and the Botanic Gardens came under the charge of Dr W. Montgomerie, an Assistant Surgeon of the Bengal Medical Corp. A resident of the island since 1819, Montgomerie was known more for his horticultural skills than his medical achievements. His large estate at Woodsville raised a variety of crops, such as sugar and nutmeg. Six hundred and nineteen nutmeg trees and 308 clove trees were planted in the Botanic Gardens in 1827 but their growth was affected by the lack of manure because there was insufficient number of cattle roaming on the island![8] In any case, nutmeg planting did become a sort of mania in Singapore in the 1850s.[9]

Unfortunately, due to its unwillingness to incur extra expenses, the Indian Government did not follow up with the development of the Gardens after 1829. It also failed to provide the much-needed support for agricultural activities pursued by her officials in the Straits Settlements.[10] There was very little development and maintenance. It was not until October 1866 that the Government of India decided to transfer the management of the Gardens to the Agri-Horticultural Society which was formed by a group of Europeans in 1837. Ground operations were supervised by British gardeners who had wide experience in tropical gardening. An extensive range of plant specimens was transplanted into the Gardens. Interestingly, pilfering of plants was not uncommon as reported by Henry Murton, the Gardens superintendent in 1878:

> [S]ome of the rarest and most beautiful ferns have been stolen from the Rockery and although a reward of $25 has been offered for the apprehension of the offenders, and a public notice given that all persons found taking plants will be prosecuted, the thefts still continue, and not confined to the Rockery, but plants are often missed from the beds and borders.[11]

It was pointed out that the native gardeners were the culprits because "they are often encouraged to introduce anything new or pretty into their employer's gardens" without fear of being interrogated by their employers.[12] Many Europeans and wealthy Chinese lived in houses with an acre or two of well-tended garden; some even had orchards or small estates. By 1868 voluntary subscription to maintain the Gardens was pegged at $15 per annum and "falls upon many (there were 87 subscribers) who can hardly afford such a sum".[13] In a letter to the Acting Colonial Secretary, the Society's Honorary Secretary George Bushell reiterated that the Gardens was a popular "institution which is not only a benefit to the inhabitants of Singapore, but a source of enjoyment and instruction to the many thousands of strangers who now call at our port, few of whom leave our shores without paying a visit to the Gardens…"[14] Even with a government grant of $50 per annum, increasing to $1,200 per annum in 1874, the

Society was in the red and not able to carry out the roles of the Botanical Gardens. Eventually, in November 1874, the running of the Singapore Botanic Gardens was transferred to the Straits Settlement government.[15] The fortune of the Botanic Gardens then became inextricably linked with the name "Mad" Henry Ridley, the Director of the Gardens in 1888 and the perpetuator of the rubber industry in Singapore and Malaya.

RUBBER AND RIDLEY

British planters experimented with sugar and coffee, but it was rubber which transformed their involvement in agriculture in Malaya and Singapore. Rubber cultivation took off at the end of the nineteenth century due to several key factors: Malaya's tropical climate, the coffee blight, world demand, the enthusiasm of Henry Ridley (Director of the Singapore Botanical Gardens, 1888–1911), the availability of cheap Tamil labour from South India, and European capital investment in modern technology.

Gutta percha, the latex of the Palaquium tree that grows in Southeast Asia and the East Indies, was widely used in handles of knives by the indigenous people. It has properties of natural plastic, tough, elastic, impervious to water and can be moulded to any shape.[16] In 1843, Dr W. Montgomerie recognized its usefulness for surgical purposes and sent some specimens to the Bengal Medical Board. Subsequently, in 1845, he again despatched some specimens but this time to the Royal Society of Arts in London. During one of the Society's meetings on 19 March 1845, Michael Faraday demonstrated the properties of gutta percha and recommended its use as an insulator.[17] The Society became the first to obtain and disseminate information about gutta percha and before long the raw material became a commercialized product used in the production of submarine telegraph cables.[18] British industrialists were quick to reap the benefits. One early pioneer was Thomas Hancock who formed the Gutta Percha Company of London in 1854 to manufacture golf balls, galoshes and tubing.[19] Following his gold medal award in 1842 for the cultivation of nutmegs in Singapore, Montgomerie was given another gold medal by the Society for his connection with gutta percha.[20] As for Singapore, the main benefactors were none other than the merchants because the price of the product skyrocketed, from $8 per picul in 1844 to $60 a picul in 1853.[21] By 1901, the price was at an incredulous $700 a picul.

It can be said that the actual phase of development of colonial science in Singapore took place with the introduction of one plant — *Hevea Braziliensis* or rubber trees.[22] In 1876, encouraged by the support of Joseph Hooker, Director of Kew Gardens, Henry Wickham smuggled out about 70,000 *Hevea* seeds from Brazil. They were shipped immediately and reached Kew on 14 June 1876. The seeds were sown the next day and 2,700 of them subsequently germinated. Ceylon was then selected as the centre of propagation, experimentation, and distribution of about 2,000 plants which were received in August of the same year. From Ceylon twenty-two plants were sent to the Botanic Gardens in Singapore. In Singapore the responsibility of achieving the economic aims of plant transfer fell into the hands of a few individual agents of change. These were the "Kew men", colonial scientists and botanists schooled and trained in Kew Gardens.

As mentioned earlier, institutional changes took place in 1874 when the administration of the Botanic Gardens was handed over to the government and, at the same time, the Raffles Museum was also set up. A qualified superintendent with a strong background in botany was sought. On the recommendation of Joseph Hooker, Henry James Murton, the first of the many Kew-trained men who devoted years of their lives to the development of the Gardens, was appointed superintendent at the young age of twenty. Murton was one of the many diasporas of Kew men sent out of England during the age of new imperialism. The typical Kew man was described by one of Kew's curators:

> In the industrial development of British colonies and possessions, the Kew man has always been among the earliest workers. As soon as the Pax Britannica has been established, and often before, he appears. He founds botanic stations where useful plants are grown for distribution, and he gives demonstrations of the best methods of cultivating them. He fostered the tea industry in India and Ceylon; he also started the cultivation of cinchona there ... Often he suffers the fate common to pioneers: he sows that others may reap. Many a Kew man has laid down his life in the conscientious performance of his duty, as genuine sacrifice to the cause of the empire and humanity as any soldier or missionary has ever made.[23]

As an agent of change, Murton brought with him many plants from Ceylon, and with later supplies from Kew, Mauritius and Brisbane, the former Agri-Horticultural Society's Park was at last converted into a proper Botanic Gardens of Singapore. He encouraged plant exchange with other botanical institutions worldwide and travelled extensively in the Malay Peninsula, thus adding new and exotic species to the Gardens' collections. His diligent investigations of economic plants led him to become the first to plant para-rubber trees in Singapore. He also published reports on native rubbers and gutta perchas. Murton left Singapore in 1879 and was replaced by another Kew-trained horticulturist, Nathaniel Cantley. Arrived as the former Assistant Superintendent of the Botanic Gardens in Mauritius, Cantley was remembered for his establishment of the forty-one hectares Economic Gardens, situated on the northern end of the Gardens. He surveyed and set up the first Forest Department and planted Malayan timber trees in the Economic Gardens. On his visit to England in 1881, Cantley took with him about 2,000 Botanical specimens collected from trees in the Singapore Botanical Gardens and jungles in the region for identification at Kew. Of these, "two proved entirely new to science — one of these, a Dracaena with ornamental foliage and of some commercial value, is being propagated with the intention of introducing it into England".[24] Cantley's zeal and hard work caused his health to break down and he died in 1887 when on leave in Tasmania.

Writing to the Secretary of State for the Colonies in April 1888 to recommend a successor to Cantley, Governor Cecil Smith reiterated that, despite the improvements made to the Gardens by the Kew-trained botanists, there remained "an immense and almost unworked field for botanical research [and] apart from matters of purely scientific interest, there is the all important subject of the cultivation of economic products to be studied, and the practical results of such study to be developed for the public benefit".[25] The Governor's stand was supported by Thiselton Dyer, the Director of the Royal Gardens at Kew who felt that the "demands upon the Superintendent of the Singapore Botanic Gardens exceeded the capacity of the kind of man which the Colony has hitherto been content to provide for the post [and] that the time has now come for the Colony to provide itself with a scientific botanical adviser of a wholly different type".[26] It did not take long for Dyer to pick the right man for the job — Henry Nicholas Ridley. He described the successor to Cantley:

> This gentleman is 33 years of age, and has been known to me for some years as a Botanist of much capacity and promise. He is an M.A. of Exeter College, Oxford, and obtained a second class in the Natural Science School on taking his B.A. Degree. He also, in 1879, obtained the Burdett-Coutts University Scholarship in Geology. Mr. Ridley is at present a

junior Assistant in the Botanical Department of the British Museum (Natural History) … Besides a wide acquaintance with systematic Botany, Mr Ridley acquired, under the late Professor Rollerston and Professor Lankester, a considerable knowledge of general Zoology. He is also fairly well versed in Entomology, an acquirement which is of great advantage to a Botanical Officer who is constantly called upon to advise as regards the ravages of insects upon cultivated crops and the best mode of combating them. Mr Ridley is not unacquainted with the conditions of life in a tropical country, as he has spent some time in Brazil, and has conducted a scientific mission to explore the Island of Fernando Noronha.[27]

H.N. Ridley took over the administration of the Gardens in November 1888 and for the next twenty-three years he worked tirelessly to make his term the most productive period in the history of Singapore Botanic Gardens. Indeed, out of the 700 Kew-trained botanists and gardeners at the end of the nineteenth century, Ridley's work has had the most lasting significance.[28] He explored much of the Malayan and Indonesian Archipelago, adding to the collections of the herbarium of thousands of hitherto unknown higher plants and ferns. He started publication of *The Agricultural Bulletin of the Malay Peninsula* and *The Agricultural Bulletin of the Straits and Federated Malay States*, which became an effective medium of knowledge diffusion. In it, Ridley contributed hundreds of scientific papers on local plants such as timbers, rattans, fibres, and drug and dye plants.

Ridley's appointment marked the development of rubber as the most outstanding plantation crop in all of British-controlled Southeast Asia. With a small budget, Ridley started an intensive course of experimentation with the *Hevea* plants. In an earlier report by the Gardens Superintendent, Nathaniel Cantley, it was highlighted that the Central American *Castilloa elastiea* and the Panama *Hevea brasiliensis* "have baffled all attempts to strike by cuttings [and] cuttings being so unsuccessful, it was resolved not to retard the trees by further cutting, but by encouraging them to grow quickly, and look forward to the production of seed".[29] Eventually, Ridley devised a method of tapping called the "herring bone," in which a thin slice of bark is pared off the edge of the initial cut, yielding more and better latex without damaging the tree.[30] His experiments on tapping were assisted by Leonard Wray, curator of the Taiping Museum and also the person taking charge of the Experimental Gardens. The "herring bone" technique remains fairly unchanged to this day. Ridley also discovered that the trees could be tapped as early as four to seven years of age, instead of the twenty-five that had been assumed. Ridley made two further innovations. From a pruning knife, chisel and mallet, he and his workers switched to the use of a modified gouge knife which is still commonly used. He also improved on the transportation of rubber seeds by packing them in containers of slightly damped charcoal powder.[31] This method is also practised today. Ridley was nicknamed "Mad Ridley" or "Rubber Ridley" because "it was his practice to stuff seeds into the pockets of planters and other begging them to make a trial".[32]

Unfortunately, the Governor of the Straits Settlements did not share Ridley's enthusiasm. Here was an instance where a conflict of interests and beliefs existed between the colonial politician and the economic botanist. Both were eager to achieve their aspirations as imperial representatives in the colonies. True to the traditions of Kew Gardens, Ridley persisted in his work on rubber. Eventually, his vision was realized when a few external developments took place which encouraged planters and officials to try planting rubber as a new plantation crop. The first was the important discovery made by John Parkins at the Botanic Gardens in Ceylon of the acid method of coagulation, producing clean rubber. This replaced the uncomfortable smoke process. It also introduced a kind of division of labour because the latex gatherers no longer did the initial processing. Another event was the collapse of the dominant coffee industry in the Malay States because of Brazilian competition and spreading disease. In desperation some coffee planters resorted to rubber. Finally, demand from the West for rubber as a raw material was rapidly rising because of the arrival of the motor age. The rubber industry of Malaya had arrived, thanks to the pioneering work of Henry Nicholas Ridley.

The booms of 1905 and 1910 in the world demand for rubber resulted in, first, more Europeans working and investing in rubber estates in Malaya and Singapore and, second, the formation of numerous United Kingdom plantation companies. J.B. Carruthers, the Director of Agriculture and Government Botanist, declared that in 1900 there were "a very small number of rubber trees, and only one or two small estates systematically planted" but by the end of 1906, more than 85,000 acres of land in 254 estates were devoted to rubber planting.[33] He added:

> As an interesting and profitable profession for a strong and healthy young Britisher, rubber planting may be confidently recommended. The life is

hard, the climate is not healthy, but by no means dangerous; there is no lack of interest in the planter's life, and the salaries earned are in most cases liberal. A man of a few years' experience can command a salary of 500 pounds or upwards, and has often opportunities of using this savings to open up either himself or with others rubber land of his own.[34]

Many of the plantation companies raised capital on the London Stock Exchange, but did so under the auspices of Singapore agency houses in Singapore (such as Guthries, Bousteads, Sime Darby, and Harrisons and Crosfield) which, with the benefit of local knowledge of the industry, supervised their activities in Malaya. Though these Singapore-based agencies were accountable to shareholders in Britain, their operations were essentially centred upon the region and, by and large, they ploughed back locally generated profits. The expansion of the rubber industry in Malaya resulted in a shift of scientific research in agri-horticulture and forestry from Singapore to Kuala Lumpur, the capital of the Federated Malay States, in the 1920s.[35] In 1925, the Rubber Research Institute of Malaya was set up in Kuala Lumpur. As explained by H.M. Burkill, the son of the former Director of Botanic Gardens in the Straits Settlements, the shift was a response to the importance of Malaya as one of the major world suppliers of rubber:

> Rubber was a boom crop. Oil palm offered high expectations. Tea and cotton were under consideration. Coffee and sugar were already established. Many diverse exotics were being brought in for trials. Furthermore, the potential of what lay in the Malayan forests was only just beginning to be realised.... It was, therefore, natural that botanical research, conducted heretofore from Singapore but intended for the economic benefit of the whole Peninsula, should follow the commercial activity in need of its to Kuala Lumpur.[36]

The Singapore Botanic Gardens continued to provide supporting services but it slipped away in importance during the remaining years of colonial rule.

VENTURES OF EUROPEAN AND CHINESE AGRICULTURISTS

In contrast with the feeble effort of the government, private individuals in Singapore were more ready to participate in agricultural ventures and research. Here, too, the Europeans took the lead. One of the earliest pioneers was Dr Jose d'Almeida who was born in Malacca but settled in Singapore. As an enterprising agriculturist, d'Almeida risked much of his wealth to experiment with new products. Cotton seeds from North America, Brazil and Egypt, cochineal, vanilla and gamboge trees from Siam, were brought into Singapore and he had them planted in his large plantation.[37] Unfortunately, these new seeds and plants failed to develop into healthy plantations. Singapore's soil and climate did not prove conducive despite a period of acclimatization. Cotton, for example, which did not need a fertile soil or expensive machinery, failed miserably because of caterpillars and frequent rain which destroyed the pods.[38]

Singapore also had a relatively modernized sugar plantation owned by another agricultural enthusiast, the American consul, James Balestier. He was described as "well known among men of science in the United States ... and the first person who has attempted the cultivation of sugar at Singapore, and for his success he was awarded the gold medal of the Calcutta Agricultural Society".[39] In the nineteenth century, sugar was increasingly demanded by an affluent Western clientele, especially the tea drinkers. It had been sought by Europeans for centuries and eagerly promoted by investors and mercantilist governments who stood to gain from increased consumption because it can be physically addictive — what Sidney Mintz calls "the drug foods" (the others are cocoa, tobacco, coffee and tea).[40] Some observers had noted that the rise of the British Empire "had less to do with the Protestant work ethic or English individualism than with the British sweet tooth".[41] By the eighteenth century, sugar production — particularly in the West Indies — had created a new model of industrial organization. Making sugar involved a series of processes, namely, cutting the sugarcane, pressing it to extract the juice, and then cooling it to allow the crystals of sugar to form. The leftover molasses was distilled into rum. The economic importance of sugar motivated plantation owners to adopt a continuous, production-line process, with labour-saving machinery and a division of work teams that specialized in separate parts of the production process. Although the Balestier Plantation had all the features of an efficient production system (which produced five to six thousand pounds of sugar per day), it was put up for sale in an advertisement in the *Singapore Free Press*, April 1848:

> The soil is good and produces on an average from twenty to twenty-five piculs of raw sugar per acre.... The buildings consist of ... a boiling house with a set of flat bottom pans — two of thick copper and three of iron — all connected and communicating with one another by means of valves, copper skimmers, filterers, etc,... An engine house and a ten-horse steam engine from the Low Moor Factory and a horizontal iron mill for crushing the cane, all in excellent working order.... And a distillery consisting of copper perpetual stills, Baglioni's patent, and fermentation vats all in working condition ... The estate is stocked with two Sydney horses and a young elephant used in ploughing....[42]

The steam engine was still a rarity in Singapore at this time, and its use in Balestier's plant allowed the operation of faster and more efficient sets of shredders and rollers through which cane stalks passed several times. Balestier did not specifically explain why he put up his plant for sale in his article in the *Logan's Journal*.[43] Perhaps the lack of support from the government in terms of capital and expertise and the overwhelmingly superior productive sugar plantations in Java proved too much for local planters to sustain the exploits in sugarcane cultivation. Sugar was an economic crop which the Dutch in Java developed to a great extent using the latest technology and scientific methods available at that time.[44] By the 1860s, however, sugar-planting in Singapore faded away because "it was decided by the Imperial Government to admit the sugar and rum of Province Wellesley into the home market at the reduced colonial duties, while the same products of Singapore was to be charged foreign duties".[45] At its peak sugar cultivation in Singapore covered no more than 400 acres.[46]

While the Europeans attempted to apply scientific knowledge to their wide variety of agricultural pursuits, the Chinese were more haphazard in their methods and restricted their interests to the plantation of sugar, gambier (*uncaria gambir*) and pepper. In Province Wellesley and Singapore, large tracts of land were taken up by the Chinese for the cultivation of sugar. Unlike the European sugar enterprises which possessed modern milling machinery, the local sugar cultivators adopted a primitive manufacturing process. John Cameron described the extracting process of the Chinese:

> Their mill consisted generally of a pair of vertical rollers, either of stone or of very hard wood, resting on a sort of bason platform raised at the rim and having an outlet leading to a large barrel sunk in the ground, and which collected the juices of the canes; the rollers were set in motion

by bullocks and the canes passed between them by the hand. Close to the mill and under the same shed a large fire-place was built of mud or mortar, with three separate shallow boilers embedded in it.[47]

Gambier was a major exchange commodity in the rice trade between the Riau Islands and Java. From Java, it was then transported to China where it was used as a tanning agent.[48] A typical large plantation had three acres and thirty acres of gambier, reflecting the economic importance of the latter.[49] On the average nine to ten men were employed, each receiving about $3 per month, depending on the market price of gambier.[50] The traditional gambier factory was rudimentary and consisted of a large attap house, with wooden walls and roof of nipalm leaves, dug-up furnaces of iron cauldrons, and other wooden implements.[51] The constant flow of immigrants from China was the main source of cheap labour. These unskilled Chinese, together with the Chinese shopkeepers in town who supplied the initial capital to those who were willing to start their own plantations, were the agents that kept the gambier and pepper industry thriving. As reported in the *Singapore Free Press* in March 1839:

> Nearly all these plantations were commenced by individuals without capital of their own, who began on small advances from the Chinese shopkeepers in town, on the security of a mortgage of their ground … the advances sometimes running on at a very high rate of interest and often made in cloth and provisions at higher than the market rates…[52]

By the end of the 1830s, there were about 350 plantations employing a large number of workers.[53] Though hampered by the lack of accessible roads, the Chinese cultivators were largely responsible for the opening up of the interior of Singapore. By 1845, about seven miles long of roads were extended to Bukit Timah, Serangoon and to the Johore Straits, providing useful access to the northern part of the island.[54] In 1849, it was reported that gambier and pepper cultivation accounted for about 60 per cent of the gross revenue in the sales of agricultural products.[55]

Beginning in the 1850s, the gambier and pepper industry began to move across the Straits to Johor. This was not due to competition of any sort but because of the wasteful method of cultivation. Large tracts of the jungle were cleared for firewood and the soil in the plantations was repeatedly used to the point of exhaustion. When no manure was added to the plot, it was abandoned and the planters moved further inland. Contemporary observers commented on the Chinese exploitation of the land:

> The Chinese undertook the growth of gambier and pepper, and gradually have extended themselves over a considerable portion of the island. But they are evil doers rather than doers of good to the land, which after a few years of cultivation they abandon, empoverished and overrun by lallang grass, and remove to a fresh clearing in the jungle, where the virgin soil becomes in its turn exhausted and a nuisance.[56]
>
> The Chinese have been the chief cultivators of gambier and pepper, but then they have no attachment to the soil. Their sole object is to scourge the land for a given time, and when worn out to leave it a desert.[57]

The Malay chiefs welcomed them and, once again, the Chinese took to the task of opening up the frontier of a vast expanse of unexplored land. In the meantime, however, the prosperity and future of the gambier and pepper industry in Singapore ended. The decline started in the 1890s with the trade depression which took a toll on local cash crop producers. Since then, diminishing demand and poor prices eventually signalled the demise of the gambier and pepper industry by the end of the First World War in 1918. The initial euphoria of Singapore's agricultural potential was short-lived when it became clear that by the 1850s the island never developed into a fertile centre for tropical produce.[58] The unfruitful agricultural pursuits of the Europeans and the indifferent attitude of the Chinese planters to scientific husbandry produced a picture of gloom to the colonial officials in Calcutta. Diffusion of any useful knowledge in agricultural science from the Europeans to the Chinese was unheard of at this time. Language was a major barrier but more important was the fact that the mainly unskilled Chinese immigrants who took up agricultural pursuits were interested in making quick economic returns and were not bothered with anything experimental. Furthermore, they had to borrow to start their ventures and certainly did not possess the capital to invest in machinery of any sort.

Despite low prices of the product, pineapple manufacturing, however, was one industry which attempted to keep up with the changing time. Interestingly, the common fruit in this part of the world was seen as a symbol of kingly wealth and power and also as a reminder of England's rise as a maritime trading power, and of its ascendancy in the West Indies in particular. In the seventeenth century England, the pineapple was known as the "fruit of kings" as King Charles II was recorded to have received one from Barbados planters and became fascinated by the fruit with its leafy crown that adorns it.[59] With the invention of the heated greenhouse in the 1680s, England's horticulturists were able to ripen the tropical young pineapples.[60] In tropical Southeast Asia, by the end of the nineteenth century, "[l]arge establishments for the tinning of pineapples exist in Singapore, where this industry first originated, and which is still the chief source of supply to the European and Australian markets".[61] The pineapple industry was largely controlled by the Chinese. The number of pineapple factories increased from ten in 1895 to twenty-two in 1897, and 200,000 cases exported in 1896.[62] In 1905, no fewer than 548,000 cases, valued at $2.75 million dollars were exported. Its continual existence into the 1930s in the commercial port-city could be attributed to improvements in tin-canning technology. In 1930, the government gave the local growers a boost by acquiring an area of 30 acres as the "site for a Pineapple Experimental Station, where cultural and manorial experiments in pineapples will be carried out, in order to develop their production as a main crop instead of a catch crop".[63] International awareness of the product from Singapore and Malaya was heightened when "displays of this product at several major exhibitions in Great Britain during recent years have undoubtedly assisted to popularise Malayan tinned pineapples and to increase the demand for them".[64] Attention was also paid to the standards of hygiene and sanitation health in pineapple factories. The position of a Canning Officer within the Department of Agriculture was created in 1936 and his task was to "establish close contact with canners [and] render them useful advice and assistance".[65]

A mention should also be made on an activity that has provided the coastal folks of the Malay Archipelago a source of income — fisheries. For centuries, Southeast Asians lived close to oceans or rivers, and harvested the sea with their traditional seine nets dragged by a number of boats working cooperatively.[66] Several historical sources, including the writings of Marco Polo, Ma Huan and John Crawfurd, had recorded the variety, abundance and quality of the fish harvest in the region.[67] By the early decades of the twentieth century, the 21,000 odd (in 1931) Chinese and Malay fishermen had to face up with the more innovative fishing methods, such as drift netting and trolling for tunny, of the Japanese fishermen. It is difficult to pinpoint their first arrival, but by 1926 there were 340 Japanese fishermen in Singapore and the community increased to 688 by 1929. By which time, they had landed 42 per cent of the total catch of fish in Singapore and "[i]t is all iced fresh fish of good quality, which however could be improved by better methods of freezing at sea and by the provision of better marketing equipment".[68] A notable advantage they had over the Malay and Chinese counterparts was the use of diesel motor-powered vessels which allowed them to fish way beyond the coast. The Malay and Chinese fishermen "do not often venture out to sea, but, as a rule, pursue their calling in inshore waters with small craft, the most common of these being the koleh, which carries a crew of three men".[69] Consumption of fresh fish increased in the 1930s as the more affluent locals and European consumers preferred them to the canned or salted fish. To encourage local fishermen and fish dealers to increase the supply of fresh fish — hitherto supplied by the Japanese and local Chinese importers who carry fish on ice from all parts of the neighbouring Rhio Archipelago — the municipal government set aside funds to provide for a suitable vessel fitted with refrigeration to carry fish from the local fishing fleets directly to the market.

Like the Rubber Research Institute set up to apply science to economic agriculture, by the 1930s, science and technology was gradually applied to the Malayan fishing industry in various aspects. As reported in 1931: "Plans have been submitted for a small biological station to enable the Department to follow up some of the most important economic fishery problems of this country, and plans for a small suitable fishing-research vessel have been submitted to the Government".[70] Experimental work in canning and salting was carried out and directed towards establishing methods of preservation on a scale suitable for the needs of small communities. Canned fish was seen to be of superior quality than the rotten salted fish. As reported to the Legislative Council:

> A considerable amount of experimental work has been done, particularly on salting. It is invariably stated that the main cause of the low grade salt fish produced in the country is due to the poor solar salt which is used, and that the employment of a high grade salt would effect a great improvement. This has proved to be incorrect, and, although possibly high grade salt may be preferable in some countries, its use in Malaya does not seem to be justified. This is due to the fact that almost all the fish which is salted have usually been allowed to decompose to a greater or less degree before salt is applied and since salt fish is largely a condiment and not a food, flavour is most important. This flavour is acquired during the subsequent bacterial processes and is compatible with the ripening of cheese. A heavily salted product is not tolerated by the bulk of the Asiatic salt fish eaters.[71]

It was found that by storing salted fish in vacuum containers or in carbonic acid gas, the lifespan of the product could increase from one to six months. Another interesting experimental work done was the estimation of the fat content in the flesh of a common food fish — a small horse mackerel. The results indicated low fat content in these fishes.[72]

APATHY TOWARDS SCIENCE

Colonial Singapore did not experience an "agricultural revolution", characterized by the development and spread of agriculture technology and farming techniques, or anything near to it. Aside from the generally infertile soil in Singapore, one can note the fact that Southeast Asia as an agricultural region has not witnessed a historical period during which the advances in agricultural technology changed the way people produced, which then increased production and motivated the search for higher skills and knowledge to sustain this gain. The mountainous mainland in the region are covered with tropical rainforests which provide the people with fruits, lumber, spice, aromatic woods, resins and home to an abundance of wild life. But rainforest soil is fragile, and once a field is cleared, it rapidly loses its fertility. Consequently, farmers adopt a form of swidden agriculture, in which the natural vegetation is cleared by cutting it down and burning it off the fields to recover its fertility and be farmed again. Anthony Reid comments that "[t]he tools of agriculture were remarkably simple and uniform, and all showed sparing use of iron which was scarce [and] for wet-rice agriculture the key items were a metal-tipped wooden plough and a wooden rake, both towed behind a buffalo or cow".[73] The climate and environment were also contributory factors. Alfred Crosby explains the general lack of development of agricultural science in the tropics: "The quantity of light in the tropics is, of course, enormous, but less than one might think, because of the cloudiness and haziness of the wet tropics and the unvarying length of the day year-round. There are no long, long summer days in the tropics. These factors, added to such matters as tropical pests, diseases, and the scarcity of fertile soil, render the torrid zone of lower agricultural potential than the temperate zone".[74] While imperial governments in the region invested in botanical gardens and agricultural research stations to facilitate the introduction of fauna and flora from other parts of the world, the technology and skills that were transforming agriculture in the temperate colonies were not applied to their tropical possessions. In most cases, successful advances in agricultural technology in Asia were based on existing indigenous traditions of invention, not on the acquisition of industrial knowledge coming out of Europe. However, it is also misleading to state that colonial science did not find its way to a colony that was largely established as a trading outpost.

Stamford Raffles himself had conducted a meticulous scientific survey of the flora and fauna and a collection of plant and animal specimens, and some three thousand drawings and maps of the region. When news of Singapore's rapid growth during the immediate years of its founding in 1819 reached London, India and surrounding lands, European botanists, naturalists, explorers, travellers, physicians, missionaries, and others began to arrive. Like the other colonized lands in Southeast Asia, Singapore attracted these trained and amateur men of science because "the region offered continual testimonies to its wealth of natural resources [and] during the period of high colonialism, plantation agriculture became the new metaphor for the region's wealth ..."[75] Thus, science during this phase is more or less an extension of geographical exploration when it is first necessary to survey, classify, and appraise the environment. Soon after the "founding" of Singapore, a series of hydrographic surveys of the island and its surrounding islets, atmospherical and astronomical observations were carried out by a steady stream of scientific practitioners, such as Nathaniel Wallich, John Crawfurd, Jose d'Almeida, Captain Franklin, Sir Harry Keppel, Charles Elliot and Jack S. Schafer (eldest son of the great British physiologist, Sir Edward Albert Sharpey Schafer). The individual's experience gained from the study of natural history and geographical sciences in a foreign land helped to modify his scientific views. More often than not, all the plants, animals, minerals and other scientific information were returned to Europe for the benefits of its homeland scientists and for stocking up zoological and botanical gardens, herbariums and museums of Europe — although, in the case of Singapore, the Raffles Library and Museum (established in 1874) had a collection of some 30,000 volumes of literary works and the museum collections related to the zoology, botany, geology, ethnology, and numismatics of the Malay Peninsula. Mention was also made above of Henry Ridley's contribution to the rubber plantation industry. His discovery of repeating the same cut and of the renewal of the cut bark was seen as both a scientific as well as an economic achievement. Johaness Dijkman equates Ridley's achievement to the invention of the pneumatic tyre and vulcanization process.[76] In short, the exploratory researches of botanists, geographers and surveyors contributed to the production of new knowledge in botany and geography and to the emergence of botany as a modern scientific discipline in the trading port-city. In the process, as described by Zaheer Baber of the situation in India, these individuals were "engaged in delicate balancing acts between their interest in botany and their attempts at seeking patronage for their researchers by articulating their scientific knowledge within the context of the imperatives of colonialism".[77]

The late 1920s was also a period which saw the gathering of support in London of the need to put agricultural development in the British colonial empire on a more scientific footing.[78] Lobbyists, such as Frank Leonard Engledow, the Chair of the Cambridge School of Agriculture, believed that the Colonial Office had to adopt a more proactive interventionist policy towards managing the Empire's natural assets which were wasted away by unscientific and traditional agricultural methods used by the indigenous populations. The impetus towards a "science for development" ideology and reflected in a concerted effort in nurturing scientific agricultural research in the colonies was provided by politicians like Leopold Amery and William Ormsby-Gore, the Colonial Secretary and Under-secretary respectively from 1924 to 1929. While this new colonial initiative would create an immense demand for new kinds of knowledge and expertise, the underpinning motivation was "the fear of colonial disorder brought on by ecological degradation and an impending 'population crisis'".[79] Experts and politicians, in the 1930s, in London persistently warned of the dangers of exploding, surplus population in the colonies feeding on limited agricultural resources, a situation that would result in heavy loss of productive resources. In turn, this would create severe repercussions for British manufacturers.

The call was made from the British Colonial Office for the setting up research stations overseas, whose activities, including contributing funding support, would be coordinated by the proposed Imperial College of Tropical Agriculture, a central research organization in London. However, there was resistance by colonial administrations, especially from the richer colonies in the East Indies, such as Ceylon and the Straits Settlements.[80] There was also the strong objection towards the imposition of imperial control on the local staff and contributing colonial revenues to the central fund in London where they had no say or control. These colonies had already accumulated years of scientific experimentations, as reflected in the setting up research institutions and creating their local scientific networks. The Royal Botanical Gardens at Peradeniya and Singapore, and the Rubber and Tea Research Institutes that were founded in Ceylon and Malaya in the 1920s were disseminating research findings through their own scientific journals and providing training for local personnel. It was the pioneering research in the *hevea* plant in Singapore that led to the rise of Malaya's rubber industry in the 1890s and the early 1900s, although it was argued that the transfer of the bud grafting technique into Malaya — rubber research carried out in Buitenzorg and Medan in the Dutch East Indies — had also contributed tremendously to Malaya's rubber output.[81]

As the port-city continued to prosper into the twentieth century, the flurry of scientific activities involving the exploration of natural history tapered off. Except for the economic botanists from Kew and professionals working for the Municipality, Western engineers and business entrepreneurs were the key individuals who were busily reaping the rewards of a successful entrepôt trade. There is hardly any record of individuals whose scientific researches had gained official attention or even may challenge or surpass the work of European savants. A contemporary observer explained the lack of a scientific culture: "The general trend of medical and scientific work in Singapore throughout the [nineteenth] century is not especially remarkable for constructive ability or statesmanlike policy. It must be remembered, however, that the handicaps were many, comprising not only popular apathy or official indifference, but also a progress in scientific knowledge which was but of slow and gradual growth".[82] The failure of colonial science to take roots in Singapore meant that the transition from the colonial science phase to what Basalla called the "independent scientific tradition" phase was extremely difficult.[83] Conditions in colonial Singapore were not conducive to the growth of such a scientific tradition. The soil was generally infertile which made research in agricultural science unrewarding. Colonial scientific education and organizations were grossly inadequate, and there were insufficient fellow scientists or researchers who could form a critical mass for reciprocal, intellectual stimulation. The scientific enthusiast was thus totally dependent on his external links with Europe to sustain his interests. Under such circumstances, a scientific culture, within the trading environment, was not germinated in Singapore.

Notes

1. Lucille Brockway, *Science and Colonial Expansion: The Role of the British Royal Botanic Gardens* (Academic Press, 1979), p. 189.

2. Quoted in Tom Standage, *An Edible History of Humanity* (New York: Walker & Company, 2009), p. 110.
3. W. Montgomerie, "Report upon the present state of the Honourable Company's Botanical Garden at Singapore, 1st February 1827", *Journal of the Malaysian Branch of the Royal Asiatic Society* XLII, Part 1 (1969): 62–65.
4. Alan Macfarlane and Iris Macfarlane, *The Empire of Tea* (Woodstock & New York: The Overlook Press, 2003), pp. 130–34.
5. For some recent works which have chapters on the history of spices and the spice trade before the nineteenth century, see William Bernstein, *A Splendid Exchange: How Trade Shaped the World* (London: Atlantic Books, 2008) and Giles Milton, *Nathaniel's Nutmeg: How One Man's Courage Changed the Course of History* (London: Sceptre, 1999).
6. Montgomerie, "Botanical Garden", p. 64. Interestingly, in his selection of Singapore as a British base, Raffles made no mention of the island's soil fertility, climate or natural resources. He was obviously more captivated by its excellent strategic location.
7. G.E. Brooke, "Botanic Gardens and Economic Notes", in *One Hundred Year of Singapore*, edited by Walter Makepeace, Gilbert E. Brooke and Roland St. J. Braddell (London: John Murray, 1921; reprinted Singapore: Oxford University Press, 1991), Vol. 2, p. 66.
8. Montgomerie, "Botanical Garden", p. 63.
9. John Cameron, *Our Tropical Possessions in Malayan India* (London: Smith Wilder, 1865; reprinted Kuala Lumpur: Oxford University Press, 1965), p. 168.
10. Buckley, *Old Times in Singapore*, p. 483.
11. Papers laid before the Legislative Council of the Straits Settlements, no. 15, 16 April 1878.
12. Ibid.
13. *Copy of Letters from the Honorary Secretary of the Agri-Horticultural Society, requesting an increased grant in aid of the Society from the public funds.* Paper laid before the Legislative Council of the Straits Settlements, 10th November 1868, p. 59.
14. Ibid.
15. *Memorandum by the Colonial Secretary as to the Gardens, Library and Museum, with reference to Despatch from the Secretary of State of 13th August 1877.* Papers laid before the Legislative Council of the Straits Settlements, 5th November 1878, p. 32.
16. H.N. Ridley, "Botany", in *Twentieth Century Impressions of British Malaya: Its History, People, Commerce, Industries, and Resources*, edited by A. Wright and H.A. Cartwright (London, 1908; reprinted Singapore: Graham Brash, 1989), p. 185.

17. Derek Hudson and Kenneth W. Luckhurst, *The Royal Society of Arts, 1754–1954* (London: John Murray, 1954), p. 142.
18. Ibid.
19. Ibid., p. 221.
20. Buckley, *Old Times in Singapore*, p. 404.
21. Ibid.
22. A model of the diffusion of Western science throughout the world and its introduction into non-European countries is prescribed by George Basalla, "The Spread of Western Science", *Science* 156 (1967): 611–22. Basalla provides an overlapping, three-stage model. Phase One is characterized by the arrival of the Europeans to non-European countries where they carried out survey of land and collection of flora and fauna. Phase Two is the period of colonial science during which a colony or country depends heavily on the scientific traditions of a technologically and scientifically advanced nation. The final phase is marked by the emergence of an independent, indigenous scientific tradition in the country.
23. Ibid., p. 192.
24. "Report by Mr Cantley on a short visit to England in 1881". Paper laid before the Legislative Council of the Straits Settlements, No. 7, 7 March 1882.
25. Paper laid before the Legislative Council of the Straits Settlements, No. 12, October 1888.
26. Ibid., C54–55.
27. Ibid., C56.
28. Ibid., C85.
29. "Annual Report of the Botanical and Zoological Gardens, Singapore, for 1881". Paper laid before the Legislative Council of the Straits Settlements, No. 7, 7 March 1882.
30. "Report on the experiment tapping of para rubber trees in the Botanic Gardens, Singapore, for the Year 1904". Paper laid before the Legislative Council of the Straits Settlements, No. 74, 22 December 1905.
31. Colin Barlow, *The Natural Rubber Industry: Its Development, Technology, and Economy in Malaysia* (Kuala Lumpur: Oxford University Press, 1978), pp. 21–22.
32. John Drabble, *Rubber in Malaya, 1876–1922: The Genesis of the Industry* (Kuala Lumpur: Oxford University Press, 1973); Song Ong Siang. *One Hundred Years' History of the Chinese in Singapore* (London: John Murray, reprinted Singapore: Oxford University Press, 1991), p. 449.
33. J.B. Carruthers, "Rubber", in *Twentieth Century Impressions*, edited by Wright and Cartwright, pp. 195–96.
34. Ibid., p. 200.

35. H.M. Burkill, "Murray Ross Henderson, 1899–1983 and some notes on the administration of botanical research in Malaya", *Journal of the Malaysian Branch of the Royal Asiatic Society* 56, Part 2 (1983).
36. Ibid.
37. Buckley, *Old Times in Singapore*, p. 185.
38. G.W. Earl, "On the Culture of Cotton in the Straits Settlements", *Journal of the Indian Archipelago and Eastern Asia* IV (1850): 720–27.
39. Charles Wilkes, *The Singapore Chapter of the Narrative of the United States Exploring Expedition* (Phildadelphia: Lea & Blanchard, 1845; reprinted Singapore: Antiques of the Orient, 1984) p. 3.
40. Sidney Mintz, *Sweetness and Power: The Place of Sugar in Modern History*. (New York: Penguin, 1985), p. 108. What people liked most about these new drugs was that offered a very different kind of stimulus from the traditional European, alcohol. Alcohol is, technically, a depressant. Glucose, caffeine and nicotine, on the other hand, were the eighteenth century equivalent of "uppers". Taken together, these new drugs gave the English society an almighty "rush" — a rush nearly every consumer can experience.
41. Ferguson, *Empire*, p. 13.
42. Quoted in Buckley, *Old Times in Singapore*, p. 483.
43. J. Balestier, "View of the State of Agriculture in the British Possessions in the State of Malacca", *Journal of the Indian Archipelago and Eastern Asia* II (1848): 139–50.
44. Headrick, *Tentacles of Progress*, pp. 240–43.
45. Cameron, *Our Tropical Possession*, p. 338.
46. Balestier, "View of the State of Agriculture", pp. 147–48.
47. Cameron, *Our Tropical Possessions*, pp. 338–39.
48. Carl Trocki, *Prince of Pirates* (Singapore: Singapore University Press, 1979), p. 19. The use of gambier in China's textile industry became significant early in the eighteenth century.
49. R.A. Jackson, *Immigrant Labour and Development of Malaya* (Federation of Malaya: Government Press, 1961), p. 22.
50. Seah Eu Chin, "The Chinese in Singapore: General Sketch of the Numbers, Tribes and Avocations of the Chinese in Singapore", *Journal of the Indian Archipelago and Eastern Asia* II (1848). Seah Eu Chin (1805–83) was the most prominent owner of gambier plantations, earning the nickname "The Gambier King".
51. Jackson, *Immigrant Labour*, p. 24.
52. *SFP*, 28 March 1839.
53. *SFP*, March 1839, quoted in Song, *One Hundred Years' History of the Chinese in Singapore*, p. 35.
54. Turnbull, *History of Singapore*, p. 47.

55. Brooke, "Botanic Gardens", p. 71.
56. Quoted in J.C. Jackson, *Planters and Speculators: Chinese and European Agricultural Enterprises in Malaya 1786–1921* (Kuala Lumpur: University of Malaya Press, 1968), p. 23.
57. Quoted in Song, *One Hundred Years' History of the Chinese*, p. 37.
58. Jackson, *Immigrant Labour*, p. 25.
59. Standage, *An Edible History of Humanity*, p. 107.
60. Ibid., p. 109.
61. *Annual Report of the Straits Settlements*, 1899, p. 16.
62. *Straits Times*, 9 August 1895, 12 January 1897 and 6 July 1900.
63. *Straits Settlements Progress Report 1930*, No. 73, 29 September 1930, C291.
64. Ibid.
65. *Straits Settlements Progress Report 1930*, No. 91, 26 October 1936, C258.
66. Anthony Reid, *Southeast Asia in the Age of Commerce, 1450–1680* (New Haven and London: Yale University Press, 1988), p. 29.
67. Ibid.
68. *Straits Settlements Progress Report 1930*, No. 73, C302, 29 September 1930, and No. 107, C479, 24 September 1934.
69. Wright and Cartwright, eds., *Twentieth Century Impressions*, p. 215.
70. *Straits Settlements Progress Report 1931*, No. 85, 28 September 1931, C323.
71. *Straits Settlements Progress Report 1934*, No. 107, 24 September 1934, C480.
72. *Straits Settlements Progress Report 1935*, No. 112, 28 October 1935, C347.
73. Reid, *Age of Commerce*, p. 26.
74. Alfred W. Crosby, *Ecological Imperialism: The Biological Expansion of Europe, 900–1900* (Cambridge: Cambridge University Press, 1986), p. 305.
75. Victor Savage, *Western Impression of Nature and Landscape in Southeast Asia* (Singapore: Singapore University Press, 1984), p. 326.
76. Johaness Dijkman, *Hevea: Thirty Years of Research in the Far East* (Coral Gables: University of Miami Press, 1952), pp. 60–61.
77. Zaheer Baber, *The Science of Empire: Scientific Knowledge, Civilization, and Colonial Rule in India* (New York: State University of New York Press, 1996), p. 174.
78. Joseph M. Hodge, "Science, Development, and Empire: The Colonial Advisory Council on Agriculture and Animal Health, 1929–43", *Journal of Imperial and Commonwealth History* 30 (2002).
79. Ibid., p. 2.
80. Ibid., pp. 5–6.

81. See K.T. Joseph, "Agricultural History of Peninsula Malaysia: Contributions from Indonesia", *JMBRAS* 81, Part 1 (2008): 7–18 and John Drabble, "A Note on Agricultural History of Peninsula Malaysia", *JMBRAS* 82, Part 1 (2009): 113–17. Professor Joseph's response to the note was included in the same article, pp. 117–19.
82. G.E. Brooke, "Medical Work and Institution", in Makepeace et al., eds., *One Hundred Year of Singapore*, Vol. 2, p. 489.
83. Basalla, "The Spread of Western Science", p. 617.

7
Food and Singapore Cold Storage

During the span of the Victorian Age, from 1837 to 1901, Britain's imperial trade had not only shifted the British economy and class system, but also indelibly changed its tastes and aesthetics. The last quarter of the nineteenth century saw an increasing mechanization of food production, processing and distribution for mass consumption and a corresponding expansion of the grocery trade in western Europe, especially Britain and the United States. The fledging canning industry in the United States was spurred by the need to supply food for troops during the Civil War. By 1870, agriculture in these countries became mechanized and more science-based. Food processing, such as canning and meat packing, was already done on a factory level. Rapid improvement in public transport technology enabled town dwellers to settle further and further into the countryside.

To cater to the new, greatly extended cities, new technology and processes of food preparation, preservation and transport had to be created. Food could no longer be brought from the farm to the doorstep. Without modern food technology, milk had to be boiled for much more than 12 hours, meats had to be cooked and salted and eggs pickled. But the urbanites of the early twentieth century industrial cities wanted to be pampered with the good things in life. By the 1880s new food and manufacturing technology was instrumental in meeting the changing tastes of consumers. It was now possible to produce standardized products in large volume and in small packages through the use of continuous process machinery. This led to the branded production of everything from toothpaste, chewing gum, photographic film to breakfast cereals, soups and canned products, all heavily promoted through newspaper and billboard advertisements. The increasing popularity of the English breakfast and afternoon tea created a rising demand for a wider range of manufactured biscuits, jams and factory-produced cakes amongst the working class living in the cities and suburban areas. Manufacturing firms, destined to be household names in the food industry, were established during this period and quickly monopolized some popular products. Peek Freans introduced the first "cream cracker" in the 1880s, and McVities gave the world the "digestive" biscuits in the 1890s. Campbell's, Heinz, Quaker Oats, Carnation, Lipton, Kellogg, Procter & Gamble were other household names that eventually come to dominate food manufacturing on a global scale. The Heinz company was already producing bottled ketchup and factory pickles, and Campbell's by this time was just about to start marketing their range of canned soups. By the start of the new century, cheese and butter making had become largely a factory operation, made easier and cheaper by the invention of the centrifugal cream separator in 1879. All in all, the availability of quality food contributed to higher standards of health and living in the industrialized West.

While the consumer society in the industrialized Britain was rapidly transformed by advancement in science and technology, the British colonies of Singapore and Malaya in the periphery of the Empire were also enjoying peace and prosperity under *Pax Britannica* and benefitting from the importation of an extensive range of consumer goods. Life in the hot and humid tropical colony was becoming more pleasant. Writing in the early 1860s, John Cameron commented on the range of food consumed by the Europeans:

> [Breakfast consisted of a] little fish, some curry and rice, and perhaps a couple of eggs… Tiffin time, a plate of curry and rice and some fruit or it may be a simple biscuit with a glass of beer or claret … [and as for dinner], soup and fish generally precede the substantials, which are a solid nature, consisting of roast beef or mutton, turkey or capon, supplemented by side-dishes of tongue, fowl, cutlets, or such like, together with an abundant supply of vegetables, including potatoes, nearly equal to English ones grown in China or India, and also cabbages from Java…[1]

By the turn of the twentieth century, the port-city of Singapore staked its claim as the world's seventh largest port in tonnage of shipping. Electricity was introduced and after 1906 electric lighting and fans replaced the manually operated oil lamps and punkahs in private houses. Several sport and luxurious social clubs were formed during the last decade of the nineteenth century and provided entertainment for the Europeans and affluent Asians. This rising level of comfort was enhanced by the incorporation of the Singapore Cold Storage in 1903. The availability of cold storage facilities and imported frozen meat, fresh butter, fruit, and other products impacted strongly on the consumer society in colonial Singapore and Malaya. The Europeans (and the more affluent local population) were now able to enjoy consumer goods imported from all over the Empire and to live a lifestyle that reflected the power and prestige of an imperial nation. Singapore Cold Storage itself was a symbol of British imperial taste.

INCORPORATING THE SINGAPORE COLD STORAGE

Writing in 1852, an observer in Singapore commented: "Considering the large European society resident in Singapore, there seems no reason why supplies of every kind should not be as abundant there as in Calcutta; but, strange to say, no good beef or mutton is to be had on the island; a small, skinny, sickly looking animal, dignified by the name of a Bengal grain fed sheep, is slaughtered twice a week for the benefit of those who cannot dispense with their mutton chop, and is sold at a fixed price of two and a half dollars the joint".[2] There were grouses, especially amongst the British and Australian inhabitants, that the local meat was too tough. In the opinion of the *Straits Times*, "it is a question whether many of these so-called sheep and goat would in reality pass inspection by the veterinary surgeon".[3] These comments indicated the dissatisfaction of the Westerners on local produce.

Out in the remote mining areas in Pahang in the Malay Peninsula, Deburgh Persse, Chairman of the Raub Australian Gold Mining Co. Ltd (with its Head Office in Queens Street, Brisbane) was also lamenting on the lack of good meat in 1900. It was also pointed out to him that the reason why Australian engineers and managers came to the mines and left soon enough was because they had to endure the consumption of local buffalo meat. This triggered Persse to explore a venture that was destined to change the food consumption pattern of the people and food-retailing scene in Singapore and Malaya. On his return to Queensland, Australia,

Persse (who was also the Chairman of the Queensland Meat Export and Agency Co. Ltd.) discussed the plight of the Australians working in the tropics with H.W.H. Stevens, the manager of Victoria River Downs, one of the world's largest cattle stations. The company had been exporting Australian bullocks from Darwin to Java and Singapore in the 1890s. But the trade was discontinued because the hot, humid conditions led to the destruction of a sizeable number (about 20 per cent) of the herd. There was only one way to restore the export of Australia meat to the tropics — the availability of cold storage facilities. The two entrepreneurs looked hard at the possibilities and decided that such a venture could succeed in Singapore where economic conditions were generally favourable. The port attracted many European ships throughout the year. At the same time, there was already a European settlement and peace and prosperity of the trading emporium was protected by the British army contingent.

Singapore Cold Storage Company Limited was registered on Monday, 8 June 1903, with its office at Battery Road and managed by H.W.H. Stevens. It started operation with a nominal capital of $600,000, and a paid-up capital of $240,000 in $10 shares of which Deburgh Persse's Queensland Meat Export took up $140,000. The enterprise was formed "for the purpose of providing cold storage, and initiating and carrying on the business of introducing, storing and distributing frozen beef, mutton, lamb, game, fresh butter, fruit and other Australian food supplies and products".[4] A piece of land at Borneo Wharf in Keppel Harbour was purchased and a cold store was constructed with two insulated rooms, each with a capacity of storing 200 tons of frozen meat. A representative of the *Straits Times* was invited by H.W.H. Stevens to visit the "Arctic region" of the Cold Storage building. As reported in the *Straits Times*:

> The cold was intense in spite of the greatcoats and wraps. The first thing that caught one's eye was the coating of frost which clung to the metal duct along which the cold air is forced into the chamber. This cold blast, it appeared, was entering the room at 12 degrees Fahrenheit. The temperature in the room itself was 26 degrees…there is also a thawing room in which frozen goods are brought back to about 32 degrees ready for the market. There are rows of overhead rails and running hooks to make the handling of the carcasses easy.[5]

The final test to the cold rooms was done by the Linde British Refrigeration Company, and by late February 1905, Cold Storage's main storage centre at Borneo Wharf was ready to start business. Another milestone in food retailing in Singapore was created when Messrs Lewis and Lambert, acting as retail agents for Singapore Cold Storage, "will have retail depots at Orchard Road, close to the railway bridge, Raffles Quay, Keppel Road near New Harbour Dock gates, the junction of Rochore Canal Road and Selegie Road, from which supplies will be delivered to private homes".[6] Retail operations started on 30 March 1905. There is no existing record of what the Orchard Road Cold Storage was like in the early decades of its inception.[7] As in many similar grocery shops in Britain and Australia during this period, the Orchard Road depot concentrated mainly on selling imported foodstuffs within the working-class district — the suburban enclaves of Orchard Road, Emerald Hill, Selegie Road and River Valley Road. Initially, the shop was basic, with no frills. An interesting description appeared in the *Straits Times* in 1910:

> The contents of this company's stall are essentially Australian, even to the ornamentation; but that is not to be wondered at as the company gets all its merchandise from the antipodes. Frozen quarters of beef and refrigerated carcasses of mutton give the stand an appetising appearance. In a centre is a stand whereon are shown tins of frozen milk and canned meats. The decorations consist of pictures of Australian sheep farms and cattle ranches and picturesque sheaves of wheat, barley, millet and other grains.[8]

In the decades to come, Cold Storage was to transform the food retailing industry, not just in terms of consumers' taste and choice but also in grocery management. The company was seen as the pioneer of a completely new industry — the self-service, corporate chain of supermarkets.

News of the establishment of the Singapore Cold Storage was received with excitement by the public. Food was a major part of the working-class budget. Especially for the middle and upper class, the consumption of certain "new" food took on definite status. Besides the prospect of reaping some economic returns by buying the shares, the *Straits Times* on 20 June 1903 reminded the public that the "mere fact of obtaining a healthy and wholesome supply of animal food, fresh butter, poultry, etc., of a quality about which there can be no doubt, is in itself a great inducement to become a shareholder".[9] It went on to say, rather philosophically, that "a rumpsteak from an Australian bullock and a pat of fresh butter at breakfast time will freshen up the inner man and make us better fit for the troubles and worries of business hours".[10] The prospect of having prime frozen meat imported from excellent stock-breeding countries like Australia was expected to lower the prices of local meat and, at the same time, offer the local residents the opportunity to enjoy superior quality meat. This was especially important to the contingent of British personnel in the army and navy, as highlighted in the report: "it is to be hoped that amongst those to reap the benefit of a better food supply, our British soldiers and sailors may be included. In the case of our troops, who have to bear the burden and heat of the day in all climates and for long periods of service, and whose health is of the utmost importance, this especially applies… Under the new process of refrigeration the meat if served out to the troops will be practically the same as that which forms the daily ration in our home barracks".[11] As it turned out, the British army and navy became the major client of Cold Storage, right till the day when a formal withdrawal of the British military presence from Singapore took place in 1971.

Singapore Cold Storage, however, did not have a dream start. Given all the hype, the Orchard Road retail agency failed to sustain its operations. A spate of letters that concerned the inaugural arrival and sale of frozen meat in colonial Singapore appeared in the *Straits Times* during the month of April and May of 1905.[12] As soon as its doors were opened for business, wild rumours of rotten meat were circulated. The talk was that the meat turned rotten on the way from Australia and "was all putrid, and the smell would kill a cat".[13] Lewis and Lambert announced on 28 April 1905 that "[o]wing to loss account we beg to notify the public that we are withdrawing from the business of Retail Agents for the Singapore Cold Storage Company".[14] The agency was taken over by Yeo Swee Hee with effect from 29 April 1905.[15] Cold Storage's "roller-coaster" historical beginning continued. It now faced the wrath of the local "boy" and "cook" who were regarded by the *Straits Times* as a "Cold Storage Danger".[16] The report commented:

> One gets a charming change of diet upon visiting Singapore now that Cold Storage comestibles are all the rage… Alas, for we Europeans, however, there is a danger ahead. Already the unspeakable "boy" and "cook" are in league against the Cold Storage scheme. Their 'squeeze pidgin' is threatened.[17] So they are buying badly, and cooking worse; they are making all sorts of complaints about the meat, throwing away as much of it as they dare. When will Europeans combine against the intolerable combines of these impudent menials?[18]

It must be noted that, even right up till the wartime emergency of 1941, European women — or *mems* — were happy to stay out of the kitchen and let the houseboy or cook to be fully responsible for the marketing and preparation of food for the family. The head servant was called a "boy" even though he might be twice the age of his employer. He ran the household and gave orders to the other servants, such as the cook, coolie and *amah*. Instructions about meals and menus went through the boy. It was up to his whims and fancy as to whom he would patronize for the household daily food needs. As it happened, "the Orchard Road Depot was the scene of a sort of free fight yesterday morning — one unfortunate Chinese cook, in the employ of a well-known Civil Engineer, being badly mauled by an employee in the service of the Cold Storage Coy., and probably more will be heard of this".[19] Indian butchers, who often combined to defraud the housewife, too were angered at the retailing of imported meat in Singapore. They saw their opportunities vanishing. The confidence of the shareholders and the general consumers in Singapore Cold Storage was further shaken when some consignments of frozen food from Brisbane, Australia, turned bad upon arrival at Borneo Wharf. One shipment in 1907 consisted of 1,500 boxes of butter, out of which 50 boxes were rancid.[20] In another shipment of fruits and vegetables from Western Australia nearly all turned rotten. H.W.H. Steven resigned in 1907 and retired to Queensland, but a few years later, he returned to Singapore and lived till the Japanese conquest of Singapore in 1942.[21] In the meantime, Australian Fred Heron arrived in Singapore in 1909 and wasted no time in putting things in order. He was appointed as the managing director of Singapore Cold Storage.[22]

Situational factors contributed to the growing demand for Cold Storage products during the early decades of the twentieth century. Administratively, in the Malay Peninsula, the Protected States of Selangor, Perak, Pahang and Negri Sembilan were formed into a Federation in 1896. This allowed the British to tighten their control and, at the same time, encourage investment in the Malay Peninsula, most of which came through Singapore. The hinterland of the Malay Peninsula was undergoing a swift transformation. The rubber manufacturing industry was enjoying its boom years as demand for Malayan rubber increased worldwide. This led to the opening up of forested land to develop new rubber estates. There was rapid development of the country's basic infrastructure, such as roads, railways and telegraph facilities. By 1906, some 350 miles of railway lines served major towns and cities along the western coast of the Malay Peninsula. With the completion of the Johore railway in 1909 a direct rail service from Johore to as far north as Prai in Province Wellesley became available.[23] Refrigerated railway vans came into use in moving supplies of fresh foods. Communication and quick access to Johore was also improved when the Causeway linking Singapore and the hinterland of Malaya was completed in 1923. Finally, the standard of living of many European planters and merchants had improved markedly since the start of the new century. According to a government report, the typical plantation manager "now had a higher income, a much larger-bungalow, and two or three kinds of liquour in his cabinet and could afford to buy a motor-cycle or car".[24]

The Great Depression, however, affected severely the entrepôt trade and the tin and rubber industries in Singapore and Malaya. European families had to adjust their lifestyles to the realities of the time. During this period, it was common to see articles in the newspapers and magazines advising families on "economical housekeeping". The popular magazine, *The Planter*, produced instructional information on how to double the life of a bath towel, how to revamp frocks using the skirt of one and the bodice of another, how to curry casseroled rabbit or make macaroni cheese or prawn and green pea wiggle to serve on buttered toast. But most of all, the main advice was how to avoid Cold Storage as much as possible, except for a few cheap and good dishes such as breast of veal or neck of lamb.[25] It reflected just how much the company had influenced the lifestyle the Europeans who made Singapore and Malaya their home! Aside from the depression years of 1929 to 1932, Singapore and Malaya were experiencing many good years of peace and prosperity. And so did Singapore Cold Storage. It pioneered the chain store system in the grocery trade. It was a matter of applying the economic advantages of mass production to retailing. Bulk buying and redesigning the store into functional components enabled food retailers to lower their overhead costs substantially and increase their sale volumes. The 1930s saw a flurry of expansion which saw the establishment of the corporate chain stores of Cold Storage in Singapore and Malaya. The Malacca Branch was opened in 1933, the Seremban Branch in 1936 and the Cameron Highland Branch in 1939. In Singapore, extension to the Orchard Road Branch was completed in 1935 and it started local retail delivery services.

INTRODUCING "NEW" QUALITY FOOD

Bread, milk and pork — these are common staples today. But in the early years of the twentieth century, the production of local bread, milk and pork underwent some revolutionary changes, thanks to Singapore Cold Storage role in the development of the food manufacturing industry in Singapore and Malaya. At the same time, the first forty years of the twentieth century saw demand for overseas food imports increased dramatically. Dietary expectations and the changing state of nutritional knowledge, on the part of consumers and producers, gradually led to an increased consumption of primarily processed food.

For centuries a mainstay of the human diet, bread remained an important source of both protein and energy for the expatriate and local communities. In England, for much of the nineteenth century, bread was not just an important item of a meal, it *was* the meal. Up to 80 per cent of all household expenditure was spent on food, and up to 80 per cent of this amount went to the purchase of bread.[26] Laws were introduced to govern its purity and, for a period of time, a baker caught for cheating would be deported to Australia. In Singapore, bread became the product of small-scale local bakeries. The Bengalis and Chinese owned many of the local bakeries. Although local bread was inexpensive and catered to the general population, it was of indifferent quality and, more often than not, baked under poor sanitary conditions. In Kuala Lumpur, for example, three bakers — Ah Jin of Cross Street, Syed Mohamed of Java Street and Ah Sing of High Street — were found to be operating under unhealthy environment. Inspecting the Ah Sing bakery, the Superintendent of Police, H. C. Syers, commented: "Here I found bread being kneaded in a trough which apparently had not been cleaned since it was made. The table and other utensils in use were also extremely dirty, and the whole place resembled a pig-stye rather than a bakery. The drainage is bad, and no attempt appears to have been made to keep the building clear of rubbish".[27] But there were exceptions. In Kuala Lumpur, a Singhalese established "Selangor Bakery" in Java Street and proclaimed that his bread was "made according to the European system".[28] In Singapore, one of the best known local bakeries was the Royal Bakery at Geylang, owned by a Chinese family. Although it had the advantage of modern machinery, what the company lacked was people with business acumen. Royal Bakery suffered losses, and Fred Heron was approached for assistance. The opportunity for the general manager of Cold Storage was too good to miss and, with the war behind, the company continued in earnest its "Made in Singapore" policy. In 1930, Cold Storage acquired the local bakery and developed it into a modern and viable enterprise. Under the skillful hands of two bakers from Scotland and utilizing good quality Australian flour, Cold Storage soon produced bread that was distributed throughout Singapore and Malaya. Hovis bread became a popular item because of its nutritional value.[29]

The history of milk in nineteenth century Singapore was patchy. Indeed, milk was not part of the Malayan diet. It was an "imperial" product introduced into colonial Singapore in the early twentieth century. It contributed to the proper nourishment of the Westerners living in the sultry tropics. The value and habit of milk drinking soon diffused into local wealthier families. However, besides imported milk, the fresh milk supply of Malaya came from small dairies, with nondescript mixtures of Indian breeds and Murra buffaloes, and chiefly under the control of the Bengalis community. At the turn of the new century, a report in the *Straits Times* in 1903 highlighted the potential health hazard:

> To everybody who chances to travel along Jalan Besar during the day, two facts are obvious. One is that all the town garbage is deposited there, the other is that it is the common feeding ground for all the milch cows belonging to the Asiatic dairymen of the town. Two questions are repeatedly asked in connection with these facts: does this garbage diet affect the milk produced by the cows? And, assuming the reply to be in the affirmative, does such milk affect the health of the consumers who are, generally, babies and young children? At present a low kind of fever is prevalent among children hereabouts, and it is held by the unlearned that it is due to the milk used, and a murmur is abroad that the cows whose produce is supplied to the public should not be permitted to feed on the garbage deposited at Jalan Besar.[30]

The situation was bad enough to warrant the attention of the municipal authorities. In November 1905 amended regulations on milk vendors were introduced into the Quarantine and Prevention of Disease Ordinance which was passed in 1886. The changes were comprehensive. The new ordinance ensured that "No person shall after the first day of January, 1906, carry on the trade of a cow-keeper, dairyman or purveyor of milk within the Municipality of the Town of Singapore without first having registered himself…"[31] Cleanliness of operations, with good ventilation, sewer drainage and plentiful supply of pure water, were required of all milk vendors. Despite the regulations to eradicate filthy habits of the dairymen and milk sellers, in 1906, there were seventy-seven convictions of adulteration of the product. Nevertheless, local entrepreneurs tried to create a market for themselves by encouraging their wealthy and well-educated compatriots to adopt the Western habit of drinking milk. They touted milk's superior nutritional properties and digestibility. By 1949, it was estimated that about 150,000 gallons of milk were produced by Chinese and Indian cattle keepers annually in Singapore.[32] However, the quality of local milk was still not favourably regarded by the European community. A former officer in the police force, Alec Dixon, described the cows as "sad and emaciated cows of Indian origin that wandered aimlessly about the waste land and scrubby wayside verges, where they were barely nourished by the coarse and wiry buffalo grass".[33] Milk was extracted in not the most desirable environment and the product was "a thin bluish-white liquid not infrequently flavoured with cow dung and usually speckled with dead flies".[34] Local milkmen often used impure water to adulterate milk leading to cases of cholera and enteric fever. It was not a product that the expatriate community or the more affluent locals would consider consuming. In the twentieth century, imported condensed milk or powdered milk became popular with the masses. In Britain, the working class could only afford the cheaper tinned skim milk. The cheaper and low-quality version of condensed milk was sweetened skim milk in tins, with all the milk fat removed and consequently low in the fat-soluble vitamins A and D. The Netherlands were the main producer and in 1916, the Cooperative Milk Condensery Friesland produced its first condensed milk and became in leader in the diary food industry. In 1906, some six million tins of milk, valued at more than a million dollars were imported into Singapore and the "Chinese preferred to spread the preparation on their bread and biscuits in place of butter".[35] Imports of condensed or powdered milk into Malaya had totalled $13 million in 1930 and this was increased to $48 million in 1949.[36] Dixon added that "[d]uring my early years in the Colony the Singapore Cold Storage Company sought to ameliorate the situation my producing milk of good quality that might be suitable for very young children".[37] The British government was also beginning to place greater emphasis on imparting the science of nutrition and proper dieting in the colonies at the periphery. Malnutrition was a grave concern and health authorities stationed throughout the Empire were instructed to improve the nutritional standards of people, especially the young. In Singapore, concerted efforts were made to educate the locals on the importance of milk consumption.[38] Supplementary milk feeds (condensed and full cream milk) were supplied to school children who were reportedly to have benefited in terms of growth and development.

For the enterprising Fred Heron, milk production was the next project on his card and it was an effort that laid the foundation of a modern milk manufacturing industry for Cold Storage. At Cold Storage's experimental dairy farm, it was scientific husbandry at its best, at least in this part of the world.[39] The initial effort was by re-constituting Klim, the well-known brand of Dutch full-cream dehydrated whole milk powder popularly used in the tropics.[40] After mixing, the milk was chilled, bottled and sold at 50 cents a pint. It was, Dixon testified, "of excellent quality and was, no doubt, the most satisfactory substitute for fresh milk that could be produced at that time — 1928".[41] In the meantime, Cold Storage made extra efforts to upgrade the milk production. The task on hand was to tackle the problems associated with the establishment of dairy farms in the tropics. None of the indigenous grasses were palatable to European Friesian cattle. A special specimen of lush, free-growing grass, coming out of the Netherlands East Indies Agricultural Research Station at Buitenzorg in Java, was found to be comparable to the grasses of the temperate, cattle-growing lands. But the Dutch scientists protected their discovery. Ironically, it was a young Dutchman, during his visit to the research station, who claimed to have pulled up a few roots of the grass and stuffed it into his pocket while officials were inattentive. He immediately flew out of Indonesia and planted the seeds in a specially prepared plot — at Cold Storage's dairy farm.[42] The grass flourished and spread very rapidly. The supply of green fodder was there. It was now time to bring in the nucleus of Friesian herd — twelve from Holland (five of the cattle died at sea) and the same number from Australia. The black and white Friesian cattle, including the remaining pigs, were constantly kept clean by a mobile electrically operated vacuum cleaner and given daily baths containing mild disinfectants. A well-designed drainage system was already in place and water was channelled into large concrete baths. The washing pool maintained the comfort and general health of the stock, as well as helping to check fly-borne infection. The cows were then milked by hand three times daily, and the milk immediately chilled in a large refrigeration chamber. Quarantine rooms were also constructed for sick animals. To prevent the outbreak of diseases, manure was collected round the clock and barns were kept spotlessly clean. The supervision of the Cold Storage's dairy farm was given to none other than the young Dutchman, Driebergen, whose experience and veterinary knowledge allowed local hands, mainly Malays and Indians, to understand and learn basic principles of milking, feeding and washing the animals. The incorporation of scientific knowledge and modern machinery had given the Singapore Cold Storage the confidence to describe its milk production in their advertisement in the *Straits Times* in the 1930s as: "Pure Fresh Milk: From our own Farm at Bukit Timah. Our fairy cows are the finest procurable, and they are kept in the most hygienic surroundings, free from all dirt and contamination" and the company "makes no pretence of catering for the poorer classes".[43]

In 1927, Cold Storage ventured into production of local pork. Hitherto, virtually all of Singapore's pork was imported from China or from Bali in Indonesia. The pork import trade suffered the risk of disease because of the lack of refrigeration facilities during freight and the lack of strict government health regulations. Moreover, the local pork trade was largely monopolized by Chinese butchers who formed a tightly-controlled organization to enforce price and supply in the lucrative business. To circumvent these problems, Fred Heron decided that Cold Storage could grow its own pigs within an environment of high standard of hygiene. This would be his selling point as local consumers would then have the ease of mind to purchase quality pork from Cold Storage. A piece of land was first rented along Bukit Timah Road, near to the present Jalan Jurong Kechil, to start an experimental station and a few large Berkshire boars were imported and crossed with local sows. Within three years there was a herd of 1,200 pigs. As the venture was deemed successful, the pigs were moved further up the road to the present Dairy Farm estate. A 60-acre stretch of forested land was purchased in 1929 and transformed into a magnificent farm that had a good drainage system which prevented the breeding of malarial mosquitoes. The pig farm, however, did not quite succeed because an epidemic of swine fever broke out in the farm in 1930 and 900 pigs died and, even before some measures could be taken to restore order, the Great Depression plunged the world into an economic abyss. These were hard times for all in Singapore, including Cold Storage, and its directors were thinking twice before embarking on further ventures suggested by Fred Heron.

Before the large-scale use of modern refrigeration facilities, one basic product that Cold Storage needed to sustain its operation was ice. Just as the air-conditioner is regarded by many today as one of the greatest inventions, the introduction and subsequent widespread use of ice into tropical land was immensely appreciated by a community that had never had anything chilled in all its history. Ice first reached Calcutta in the nineteenth century through the efforts of an American lawyer-turned-entrepreneur named Frederic Tudor. From the city of Boston, ice blocks were packed in sawdust and loaded on ships which sailed 15,000 miles in four or five months to the British tropical stronghold in Calcutta. The business risk was well taken as the first run lost only a third of the ice cargo. British colonists got to enjoy a lump of ice in the afternoon cocktail.[44] By the 1890s, the development of mechanical refrigeration (through "vapour compression") enabled ice — and hence, frozen meat and other perishables — to be shipped to all corners of the world. By this time also it was possible for entrepreneurs in the food industry to build technologically advanced ice-making facilities. Cold Storage's plan to do so was first mooted to the company's directors in 1905 but they were then not comfortable with the additional expenditure to develop such an enterprise. The situation had changed positively a few years later. In the United States, the refrigerated railway car had pushed Americans to overcome their prejudices against cold meat. The technology allowed the transport of meat and other perishables from coast to coast. Americans finally accepted that meat was no longer something that needed to be eaten right after the kill. The Chicago meat-packing industry became one of the country's booming businesses.

In 1916, Singapore Cold Storage built an ice manufacturing plant at Borneo Wharf. It was no ordinary set-up because Fred Heron had made the best use of Western technology and engineers available in Singapore. Production began on the first day of July 1916. The plant had a production capacity of 15 tons a day and, because of its location at the waterfront, ice was shipped in large quantities. Over in Kuala Lumpur, ice-making facilities were built in 1918. After the First World War in 1919, Cold Storage gradually established itself as the main manufacturer of ice in Singapore. Demand was high, especially when ice was vital to the fishing industry in Malaya. Expansion of its plant at Borneo Wharf was financed through the floating of 22,500 shares at par in 1919.[45] To facilitate Cold Storage's expansion into the Malay Peninsula, a network of ice factories were built between 1921 and 1932. After Kuala Lumpur, ice works came up in Penang, Taiping, Teluk Anson, Klang, Kampar, Seremban, Sungei Patani, Kota Baru and Kuantan. But these facilities did not just cater to Cold Storage's operations. The availability of ice was a boost to the coastal fishing industry. Its products were now able to penetrate deep into the hinterland, a development that was never considered possible until the arrival of Cold Storage's ice factories. Fishing boats obtained their requirement of ice from the factory before leaving for the fishing grounds. The larger boats were usually out for two or three days and could carry as much as five tons of ice to preserve their catches.

Taking advantage of a strong presence in ice manufacturing and refrigeration business, Cold Storage moved into the production of ice cream. A small factory was set at the Borneo Wharf in 1923 to produce "Paradise" ice cream, depicting the Bird of Paradise. To get the product off the ground, Cold Storage hired an "ice cream man" from the country that pioneered the ice cream cone, the United States. But the company's ice cream venture really took off only in 1937 when "Magnolia" became its selling brand. In that year, a deal was made with Senor Soriano, the owner of the San Miguel Brewery and a successful ice cream business in Manila. The outcome was the formation of the Cold Storage Creameries Ltd., with an initial capital of $100,000 to which the Philippine company contributed 45 per cent and Cold Storage controlled 55 per cent. San Miguel transferred technology and knowledge in ice cream manufacturing into Singapore. Technicians Luis Miranda and Ponce Enrile were recruited by Cold Storage. In the years to come, Magnolia ice cream reigned as the "king" of ice creams and, indeed, led to a legal tussle between Cold Storage and San Miguel over the ownership of the brand name.

THOSE WERE THE GOOD "IMPERIAL" DAYS

"Looking back over the 1920s and '30s", writes Margaret Shannon, "there was a special quality to the life of the European community in Malaya, and many who experienced it felt in retrospect a warm glow of remembrance and gratitude".[46] The Population Census of 1931 recorded a total of 17,768 British settlers in Singapore — a far cry from the early decades of the nineteenth century when the Europeans numbered less than a thousand.[47] Singapore of the 1920s and 1930s was decorated with many new buildings, public gardens, first-rate hotels, such as Raffles and Sea View, fine European shops such as Whiteways, Laidlaw and Robinson, the Chinese-dominated businesses along North and South Bridge Roads and the bargain hunters haunt at Change Alley. The symbol that characterized the rich was the Rovers, Buicks, Hillmans, Fords, Vauxhalls or Morris. In addition to the motor car, European families possessed a retinue of domestic servants — a cook, "boy" (to clean and run the household), *syce*, chauffeur, an *amah* or *ayah* (nursemaid), gardener, and *dhobi* (laundryman). Long-time resident

Roland Braddell called the city "a kind of tropical cross between Manchester and Liverpool", while Bruce Lockhart branded it as "Liverpool with a Chinese Manchester, and Birmingham tacked on to it".[48] Marine engineer, Leslie Froggart described Singapore as one of the "sweetest smelling spots" he had visited, where "a mixture of garlic and temple flowers, durian and incense, teak wood swelling in the river and Yardley's talcum powder so popular with the Indians" pervaded the air.[49] Froggart recorded: "There's no denying it, it was easy life [in Singapore], and a pleasant one. Food was varied, plentiful and cheap. Liquor and cigarettes came in practically duty free, so that we drank whisky and smoked and thought nothing of it".[50] Mining engineer, Norman Cleveland recalled that in days when the refrigerator was not around, bottles of beers were chilled by putting them into the Shanghai jar which kept the water cold due to evaporation through its porous clay.[51] Life was generally good and comfortable for the few thousands of Europeans living in tropical Singapore.

In many ways, Cold Storage contributed to the "special quality" of everyday lifestyle. As the only hygienic European-run store in Singapore and other major towns in Malay Peninsula, Cold Storage offered good frozen meat and a variety of products from Australia, New Zealand and other parts of the industrialized world. The company came to occupy a central place in food shopping, particularly as the city-port developed its suburban areas. Never before had the European and local community enjoyed such a wide range of food products. Well-known English and Scottish proprietary brands, such as bottled sauces, pickles, jams, jellies, custards, confectionery, etc., wines from France, tomatoes from Australia, vodka from Poland, chocolates from Switzerland, Droste chocolates from Holland, brandy from South Africa and Australia, grapes from Italy, and cork (for insulation of cold rooms) from Spain, and meat and butter from the farms of New Zealand arrived at their doorstep — courtesy of Cold Storage. Cheese alone, the expatriate customer was spoilt for choice: Dutch Edam, Italian Gorgonzola, Danish Roquefort, Swiss Gruyere, English Cheddar and Australian Cheddar. Cabbages, carrots, green beans, leeks, parsnips, etc., came not just from local farms in Singapore, but were also imported from the Cameron Highlands of Malaya. Bill Goode, who served in the Malayan Civil Service in the east coast of Malaya in the 1930s recalled how Cold Storage played a crucial part in making life there more bearable: "The weekly steamers from Singapore provided a vital link with the outside world.... real meat, packed originally in ice — which by then had melted — and sawdust, all of which had to be washed off, but which made a welcome change from local goat and Chinese pork; imported apples from New Zealand and such luxuries as Iceberg Butter, which came in tin from Singapore and in the heat you had to put it on with a paint brush".[52] To cater to the daily *stengahs* (drinking sessions of soda and whisky) of the European community and the beverage needs of clubs, hotels, and restaurants, Cold Storage imported the widest range of liquor available at the Orchard Road retail store. The scale of its stock can be seen from the fact that 986 cases of liquor valued at $49,707 were destroyed on 13 and 14 February 1942, as a response to the government war order.[53] It is also interesting to note that the food retailer also stocked a range of imported perfumery — patent medicines, tonics, laxatives, aspirins, perfumes, lipsticks, soaps, antiseptics, bandages, etc. — which attracted European housewives.[54] Cold Storage was gradually becoming a one-stop shopping place — and, to a large extent, an agency for propagating an imperial culture of "Britishness".

In advertising, the imperial theme offered distinction, credibility and a sense of the exotic. Cold Storage had interesting advertisements in the *Straits Times*, selling its range of imported foodstuff, with an emphasis on the exotic, the luxurious, the convenient and the economical. Its consumer goods came from all over the world and reminded the British living in Singapore of their worldwide status. In one advertisement that appeared in November 1936, a native young man was about to dive into the river with a crocodile lurking in the water (see Figure 7.1). It explained the meaning of the caricature: "That is a capital motto suitable to many occasions. But we leap at the opportunity of reminding you to look at our stock of selected fish, before you arrange another menu. A complete in-season assortment of sea fresh halibut, blue cod, salmon, etc".[55] In another advertisement (see Figure 7.2), a caricature of a frustrated man, desperately trying to cut his plate of steak and chiding his wife, appeared with the caption "Why doesn't she go to the Cold Storage?"[56]

Gone were the days when stuffed eggs, mulligatawny soup, scraggy chicken and *ikan merah* (red mullet) formed the staple menu of gala dinners, birthday parties, anniversaries, farewells, reunions and any kind of *Hari Besar* (Big Day). Cold Storage now gave the nouveaux rich Europeans and the affluent local residents an international cuisine that would make all hosts and their guests satisfied. For diners, a typical feast at Sea View Hotel, according to Lockhart, consisted of "an excellent *petite marmite, homard a l`americaine,* and roast pheasant" and complemented with

Food and Singapore Cold Storage

Figure 7.1
"Look Before You Leap!" Advertisement in the *Straits Times*, 21 November 1936.

Source: Reproduced with permission of Cold Storage Singapore (1983) Pte Ltd.

Figure 7.2
"Why doesn't she go to the Cold Storage?" Advertisement in the *Straits Times*, 18 November 1936.

Source: Reproduced with permission of Cold Storage Singapore (1983) Pte Ltd.

bread sauce and *pommes pailles*".[57] For the lady of the house who did her own cooking, "Perfection" oil cook stoves and ovens were her favourite companion. The American brand was manufactured in Cleveland, Ohio by the Perfection Stove Company since the 1910s and, even today, it is a popular heritage item to possess. It was brought into Singapore by the Standard Oil Company of New York.

European visitors to the port-city were often fascinated at the number of "jollies" or evening parties that made life in the tropics totally enjoyable. Gin-slings, *stengahs* and frothy cocktails with egg white and streaked with crème de cacao flowed freely. Celebrating his wife's birthday on 18 December 1926, a planter working in Pahang was pleased with what Cold Storage at Kuala Lumpur could supply to his appreciative guests — pheasants, *Peche Melba*, with lashings of beer, Bristol Milk sherry, gin and whisky.[58] Interestingly, there were also expatriates, like R.J. Sidney (writing in 1926), who felt that "[t]he average European arrives in this country with a fixed notion in his mind that it is necessary to eat more here even than in a cold climate, such as England [and that] fresh meat is unobtainable and that the meat from the cold storage lacks vitamins".[59] Although there were plentiful local fruits and vegetables, many Europeans still "prefer to live on devitalised food which come from the cold storage [and] it is probably on account of this that so many Europeans lose their health or become morbid, or acquire an intellect which unfits them for work in other parts of the world".[60] Even in a smaller town like Segamat in north Johore, a branch of the Singapore Cold Storage based at a large Chinese grocery store was providing the foods that enriched the lives of the Europeans living there.[61] Social and sports clubs were the focal hubs for Europeans. Here sporting activities, parties, dances or just a few *stengahs* and a game of billiards, bridge or mah-jong took place, and calories lost were restored with a rich array of "sausage and mash", "melton mowbray pie" and plenty of good cheese. Pleasure, relaxation, sport and good food had all contributed to sustainable health.

AN IMPERIAL SYMBOL

For the Singapore Cold Storage, the evolution from a grocery store to a supermarket-style of food retailing took place in the late 1930s. It had already in operation several practices that formed the foundation of a supermarket. The retail branches were also gradually transformed through the use of some fundamental store layout and design principles. For product exposure, the customer needed an unobstructed view of the store from the exterior. The shop must not only be easily identified as a grocery store but that it also contained a variety of products to be purchased. Newly designed wooden shelf cabinets called gondolas replaced old ones. Promotional items were prominently featured on open-top display tables so that customers could see and touch the merchandise. The whole ambience of the Cold Storage store was brightly illuminated to provide customers a clean and comfortable shopping environment. There was also the popular "Winter Garden Lounge" at the Orchard Road store where shoppers could enjoy cold drinks from a soda fountain, ice cream, sundaes, etc.[62] Children were also catered for at the Junior Bar. From 1935, the company started its own free, twice-a-day delivery system. A fleet of Ford vans was added and became a "moving" advertisement for the quality food and services of Cold Storage. The increasing popularity of the private motor car allowed not only families to shop "in person" but also motivated them to purchase more groceries per shopping trip. The knowledge and skill of the engineering section at Borneo Wharf was also significant in the company's expansion plan. The European engineers, under the Superintendent Engineer J.J. Innes, adapted their knowledge to local situations. Being the pioneers in the area of tropical food refrigeration technology in Singapore, they had to rely solely on their own ingenuity to maintain, repair and develop facilities at Borneo Wharf. These engineers trained and developed the local technical and engineering workforce within Cold Storage. It was not surprising that, before the start of the Second World War, the company was able to hold over $2 million worth of stocks in its cold stores.

But this is not to say that Cold Storage had no competition of any sort. By the 1930s, enterprising Chinese shopkeepers and provision stores were responding to changing consumer needs and purchasing habits. Electricity now lighted up most parts of Singapore and the refrigerator became a common gadget in local households that could afford it. Shopkeepers themselves had to adapt and change to keep up with the times. They now stocked their supplies with more quality imported food items and initiated a home delivery system. Khoo Teng Soon, who was hardly seventeen when he joined Cold Storage as a ledger clerk stationed at Empire Dock, confirmed that, while Cold Storage was the only sort of supermarket in late 1930s, "there were small rivals, just facing the Cold Storage at Orchard Road".[63] These "European style" local provision shops, like Kim Ann, Lim Khoon Heng, Chew Nam Seng, Chin Bee and, along Robinson Road, Wah Hin, were also selling mainly to expatriates and "would stock up all the things that Cold Storage would stock, if not all but some of the more popular items, like beef, all foreign stuffs anyway".[64] Adds N. Narayana who, as a young boy, lived with his family in Cuppage Road in the 1930s:

> As money was not plentiful, a cent, or even a half-cent or quarter-cent, had some purchasing power, and to think of squandering good money at prestigious places was low on the list of the common man's priorities. Anyway local needs could easily be satisfied at less up-market local shops, at which it may have been "*infra dig*" for Caucasians to be seen shopping. The more common method of extending the life of perishables was by use of blocks of ice, which were kept in sawdust to stop from too-fast evaporation. In homes which could afford it, there were "ice boxes" which were basically wooden chests with tin or zinc lining inside where these ice blocks were put in to keep the temperature at a lower level until they melted into water.[65]

For the majority of the local population, however, food was normally bought daily at one of large wet markets at Orchard Road, Clyde Terrace, Ellenborough Street, Pasir Panjang and Telok Ayer where, "Chinese cooks with portable kitchens perambulate the streets at all hours, and distribute viands, which, however tempting to their own class, could hardly be adventured by others, since the materials of which they are composed may … be the flesh of dogs, lizards, and rats, all of which come within the scope of the Chinese cook's oracle".[66]

Although primarily established to cater to the European meat consumers, over the years Singapore Cold Storage was also becoming a household name amongst the locals because of its reputation in producing and selling some high-quality "imperial" household commodities — ice, ice-cream, bread, milk and, of course, meat. For the local families, to be seen walking into the retail store at Orchard Road and coming out with bags of "European" products was itself a kind of status symbol that reflected one's success in business and education. N. Narayana recalls:

> In those very European colonial times, ordinary Asian lesser mortals had an in-built inferiority complex which stopped most of them from stepping into high-class establishments like Cold Storage and Robinson.

Watching (with perhaps a tinge of envy) through the glass panes into the world of the upper class and better heeled was as far as those low (or lower) down the ladder could aspire to. The hawkers at the wet market supplied most of the required produce for locals, and we felt more at home haggling with them, and paying considerably less than the "white man" (did at Cold Storage).[67]

The perception that Singapore Cold Storage and its chain of supermarkets cater mostly to the Europeans and the affluent, educated locals still lingers on.

Notes

1. Cameron, *Our Tropical Possession*, pp. 295, 301–302.
2. *The Singapore Chapter of Zieke Reiziger; or Rambles in Java and the Straits in 1852 by a Bengal Civilian* (London: Simpkin, Marshall and Co.; republished, Singapore: Antiques of the Orient, 1984) p. 4.
3. *Straits Times*, 22 March 1905.
4. *Straits Times*, 20 June 1903. Refrigerated transport technology, as in refrigerated rail cars and ships, also made great strides towards the end of the nineteenth century. This allowed not only meat but other perishable fruits and vegetables to be transported over long distance. The first successful shipment of meat between Australia and England took place in 1880, when 40 tons of frozen Australian beef were landed in London. Chilled meat, as opposed to frozen, using salted ice also started to be shipped between the United States and London, as early as 1870. Improvements in refrigeration techniques made it possible to ship chilled (not frozen) meat across the Equator from about 1909.
5. *Straits Times*, 17 March 1905.
6. Ibid.
7. The retail depot started with two rented shops along Cuppage Road which were renovated with new refrigeration facilities. A few years later, two more adjoining shophouses were rented. In 1919, all four shops were purchased by Cold Storage and the foundation of the present Orchard Road Cold Storage in Singapore was laid. For an account of the business history of Singapore Cold Storage till 2003, see Goh Chor Boon, *Serving Singapore: A Hundred Years of Cold Storage, 1903–2003* (Singapore: Cold Storage Singapore, 2003).
8. *Straits Times*, 17 August 1910.
9. *Straits Times*, 20 June 1903.
10. Ibid.

Food and Singapore Cold Storage   191

11. Ibid.
12. *Straits Times*, 5, 6, 14 April, 9 and 10 May 1905.
13. *Straits Times*, 3 April 1905.
14. *Straits Times*, 28 April 1905.
15. Yeo Swee Hee was a prominent member of the Straits-born Chinese community. He was born in Singapore, being the son of Mr Yeo Kwan, and was educated at Raffles Institution. As a young man he was employed as an assistant in Huttenbach Brothers and Company (situated at 48 Short Street) since the establishment of that firm in Singapore in 1885. Gradually he expanded his economic interests into landed property, mining and general merchant and commission agency. Yeo passed away on 10 October 1909 at the age of 47. Described by the *Straits Times*, dated 12 October 1909, as a "genial type of man" his death was "generally regretted not only by his compatriots but by many European residents with whom his commercial dealings brought in contact".
16. *Straits Times*, 25 May 1905.
17. It was a common practice for the cook to "squeeze", that is, to cheat while doing the marketing of food for the household. In most cases, too, it was done in liaison with the "boy".
18. Ibid.
19. *Straits Times*, 5 April 1905.
20. K.G. Tregonning, *The Singapore Cold Storage* (Singapore: The Straits Times Press, 1969) p. 8.
21. Ibid., p. 8. The Obituaries Section of *British Malaya*, February 1945 states: "Mr H.W.H. Stevens was probably the oldest European in Singapore, as he was 91 when captured by the Japanese. Known to all affectionately as "Daddy", he was the founder of the Singapore Cold Storage and one of the founders of the Yacht Club. He died after every possible care and was buried at Bidadari Cemetery".
22. Together with George Skelton Yuill, a majority shareholder in the Queensland Meat Export and Agency Company which supplied frozen products to Singapore Cold Storage, Fred Heron knew what had to be done because Australia itself was undergoing a transformation in the nature of retailing and remaking the cultures of shopping. As elsewhere in the industrializing world, retail and consumer cultures emerged in nineteenth century Australia, giving rise to new retail forms. The early decades of the twentieth century saw the establishment of the famous household names in departmental and supermarket retailing, such as David Jones, Grace Brothers, Myer, Coles and Woolworths. The optimism of the "home experience" gave Yuill and Heron the confidence and motivation as they set out to build their own "cold storage" food retailing in Singapore. Within a year Singapore Cold

Storage increased its gross profit to $40,286, an increase of about $5,000 from the last accounts. The company also secured, for three consecutive years, the naval and military contracts for beef and mutton and, as reported in the *Straits Times* of 27 March 1910, "there has never been a complaint as to the quality of the meat, or the irregularity in delivery". Interestingly, it was a Chinese comprador, Lim Eng Teck, who helped to rejuvenate the once feeble business. Lim was employed as a shorthand typist in 1910 but due to his abilities to communicate and interact with other local shopkeepers, he was made the marketing man for the wholesale trade in food supplies, such as butter, biscuits and tinned milk. In 1965, Lim retired as a fruit and vegetables executive from Cold Storage, after having served the Company faithfully for 55 years!

23. Two of the main towns across the causeway which benefited from the immediate expansion plan of Singapore Cold Storage were Kuala Lumpur, the Federal capital, and Ipoh. In November 1909, Cumberbatch and Company was appointed as agents and a storage chamber was built in the vicinity of the Empire Hotel, a new "first-class" hotel serving "cuisine under European supervision" and situated in the heart of the town and close to the railway station. The populated town of Ipoh also welcomed the arrival of Cold Storage. In September 1909, K.A. Stevens and Fred Heron came to Ipoh and completed the arrangement. A retail and wholesale depot was erected in the Fraser and Neave's new building. It had a cold storage chamber capable of holding five tons of frozen meat and produce. Over the next few years, Cold Storage established branches in Klang, Penang, Taiping, and Telok Anson. By 1932, the year in which Fred Heron retired to begin other businesses in Western Australia, the products of Cold Storage were obtainable in most towns, as far north as Kota Baru and Kuantan on the eastern coast (both branches were opened in 1928). See *Straits Times*, 29 July 1909, 11 September 1909, 14 September 1909, 17 November 1910 and 1 December 1910.

24. J.G. Butcher, *The British in Malaya, 1880–1941: The Social History of a European Community in Colonial Southeast Asia* (Kuala Lumpur: Oxford University Press, 1979), p. 81.

25. *The Planter*, February 1932, p. 473 and May 1932, pp. 307 and 577.

26. See Christian Petersen, *Bread and the British Economy, 1770–1870* (Aldershot, England: Scolar Press, 1995).

27. Quoted in J.M. Gullick, *Kuala Lumpur, 1880–1895* (Malaysia: Pelanduk Publications, 1988), p. 73.

28. Ibid.

29. The high wheat-germ content of Hovis flour was patented by British Bakeries, one of Britain's largest bakeries, in the late 1880s. As one of Britain's biggest and longest established bread brand, Hovis began life in

1886 with the name of "Smith's Patent Germ Bread". A national competition was organized which was won by a London student who took the Latin "homminis vis" (strength of man) and shortened it to Hovis. By 1896, in Britain, there was a move to make Hovis a national product; successful intensive and extensive advertising campaigns were mounted to make Hovis a household word, synonymous with good health and good sense.

30. *Straits Times*, 30 April 1903.
31. *Amended Regulations Respecting Milk Vendors made by the Governor in Council under Section 5 (xii) of "The Quarantine and Prevention of Disease Ordinance 1886"*. Papers laid before the Legislative Council of the Straits Settlements, No. 72, 24 November 1905, C253.
32. R.A. Wright, *Report on the Veterinary Department* (Singapore: Government Printing Office, 1949), p. 5.
33. A. Dixon, "Milk Production in Singapore", *British Malaya*, March 1963, p. 28.
34. Ibid. The same situation also happened in Shanghai where the industry as a whole continually faced formidable difficulties in ensuring the cleanliness and profitability of their product. See Susan Glosser, "Milk for Health, Milk for Profit: Shanghai's Chinese Diary Industry under Japanese Occupation", in *Inventing Nanjing Road: Commercial Culture in Shanghai, 1900–1945*, edited by Sherman Cochran (New York: Cornell University, 1999), pp. 207–33.
35. *Straits Times*, 3 May 1907.
36. Wright, *Report on the Veterinary Department*, p. 5.
37. Dixon, "Milk Production", p. 28.
38. See Ng Yeen Chern, "Nutrition and Health: Milk in British Malaya", unpublished honours thesis, Department of History, National University of Singapore, 1999/2000.
39. Fred Heron was convinced that his small experimental farm could turn out to be a huge commercial enterprise if more money and land could be given. But not the Yuill trustees in Sydney. It was a wrong time for them to commit more funds. Undaunted, Fred Heron resigned from Singapore Cold Storage and, with the support of some European businessmen in Singapore, formed the Singapore Dairy Farm Ltd. to operate the farm. Rapid expansion took place. More first-class Friesian cows were imported from Canada and the United States. The buildings were air-conditioned in 1936 and, two years later, the farm staff successfully experimented with artificial insemination, producing into the world the first Singapore-born Friesian calf. Before the war broke out in 1941, 600 healthy cows were gazing proudly on a once waste land. It was a dream realized for Fred Heron who had to leave Singapore for Western Australia for health reasons in 1932.
40. "Klim" became a staple of scientific explorers in the tropics, geologists and

soldiers who needed a lightweight dry ration that would keep for several days in high heat and humidity, even when decanted from its container.
41. Ibid.
42. Ibid.
43. *Straits Times*, 28 April 1931 and 2 July 1931.
44. Fraser and Rimas, *Empires of Food*, p. 148.
45. But Cold Storage had a major competitor in the Straits Ice Company. Situated in Teck Guan Street, Straits Ice Company was started by a German concern, the Katz Brother, prior to the start of the war in 1914. As a result of the war, the British government liquidated the company in 1918. However, an enterprising manager of the former German firm, Henry Waugh, salvaged what remained of the plant and rebuilt Straits Ice Company into a modern plant with capacity of 70 tons daily. In 1926 Fred Heron successfully initiated a takeover negotiation with Henry Waugh and the Straits Ice Company came under Cold Storage for a price of $800,000. During this period too ice works were set up in major locations in Malaya where the Company was becoming a familiar food wholesaler and retailer. It is interesting to note that the acquisition strategy was to continue and contribute immensely to the Company's status, business network and market control of the food industry in Singapore.
46. Margaret Shennan, *Out in the Midday Sun: The British in Malaya 1880–1960* (London: John Murray, 2000), p. 109.
47. British Malaya, *Population Census*, 1931.
48. R. Braddell, *The Lights of Singapore* (Methuen & Co., 1934; reprinted, Kuala Lumpur: Oxford University Press, 1982), p. 113; Bruce Lockhart, *Return to Malaya* (London: Putnam, 1936), p. 140.
49. Quoted in Margaret Shennan, *Out in the Midday Sun: The British in Malaya 1880–1960* (London: John Murray, 2000), p. 113.
50. Quoted in ibid., p. 116.
51. Allen, *Plain Tales*, p. 521.
52. Quoted in ibid., pp. 582–83.
53. Personal records of W.L. Fincher, 17 September 1945. This document is from the personal files of Professor K.G. Tregonning.
54. Ibid.
55. *Straits Times*, 13 November and 21 November 1936.
56. *Straits Times*, 18 November 1936.
57. Lockhart, *Malaya*, p. 86.
58. Ibid., p. 120.
59. J.R.H. Sidney, *Malay Land* (London: Cecil Palmer, 1926), p. 86.
60. Ibid.
61. Ibid., p. 185.

62. *Straits Times*, 12 November 1936.
63. Khoo Teng Soon, *Interview*, Oral History Unit, Singapore Archives, 11 September 1984.
64. Ibid. I would also like to thank Mr Jordan Yin for his input on the local provision shops.
65. N. Narayana, Interview with author, 10 July 2002.
66. Moore and Moore, *The First 150 Years*, p. 203.
67. N. Narayana, Interview with author, 10 July 2002.

8
Politics of Imperial Education

The British administered its rule in their Asian colonies around the use of indigenous clerical and professional skills. Those who aspired to join the rank-and-file of the government services and British companies had to better themselves through acquiring an English education. In British India, English education was initially provided by missionaries, but in the 1860s, Departments of Public Instruction were created and were responsible for the development of a growing network of English-medium schools. Similarly in Hong Kong, most of the colony's seventy government-supported schools taught English by 1900. Even in Britain itself, the educational system, never considered as the most progressive of Britain's institutions, resisted the introduction of applied sciences into its university curricula. Up until the end of the nineteenth century, England had no formal institutions imparting technical education, and engineers received their training as apprentices. Indeed, India provided her colonial master with useful development models of technical institutions to replicate. The colony had established engineering colleges at Poona, Calcutta and Madras in the nineteenth century.

In the case of Singapore, the progress of education generally lacked behind the rate of economic change during the period of British rule. The ruling elite carefully steered the educational rudder so as to preserve their economic and technological status quo. The limited opportunities in English and technical education within a *laissez-faire* educational system meant that the people generally lacked the educational standards to understand Western science and technology, thus inhibiting the growth of a technological culture that could facilitate the transfer and diffusion of specialized skills and technical knowledge to the indigenous population. This is compounded by the racial prejudices of colonial administrators, who harboured the notion that Western education, including technical education, was not appropriate for the indigenous people. Instead of operating as an agent for economic and social modernization, the *laissez-faire* and often haphazard education policy in Singapore constituted a divisive force which reinforced existing communal and social barriers. Up to the 1950s, the colonial educational policy was not "national" in character and failed to promote social integration. It caused unnecessary tension between the Anglophile elite and the vernacularly-educated cohorts and further exposed the pluralistic nature of the Singapore society.

EARLY ATTEMPTS AT EDUCATIONAL CHANGES

As an accomplished botanist and inspired by the humanitarian spirit of the late eighteenth century, Stamford Raffles in 1823 established an English school for all races in the new settlement of Singapore. One salient feature of the institution was its emphasis on a scientific observation of the surrounding lands. He decreed that the science department was to teach Newtonian astronomy, mechanics, chemistry, zoology, botany and mineralogy. The teaching was to be illustrated by suitable experiments. Raffles was indeed far in advance of his time, since no government provision for education was made in England itself until 1847 and natural sciences were not generally taught in England's schools.[1] The Singapore Institution (the forerunner of today's Raffles Institution) together with St Joseph's Institution founded by the Catholic Order, provided the local inhabitants with an opportunity to be educated. In the celebrated annals of these two schools were some of Singapore's finest figures, such as Lim Boon Keng and Song Ong Siang, in the early twentieth century and several prominent political leaders of post-independent Singapore, including Lee Kuan Yew.

Despite Raffles's reminder that "education must keep pace with commerce in order that its benefits may be assured and its evil avoided", subsequent progress in educational development in nineteenth-century Singapore was painfully slow.[2] Neither the government nor the mercantile community was prepared to support the expansion of educational opportunities. The situation was succinctly stated in the local press, the

Straits Times, a few months after the control of the Straits Settlements was transferred to the Colonial Office in London in 1867:

> The duty we mean is that of every civilised and enlightened community, to use every endeavour to civilise and enlighten those with whom they are brought into contact. And how far have we performed that duty? … The future of the Straits Settlements depends in a great measure upon its present mercantile men… We must educate them properly, enabling them to compete favourably with the youth from other parts of the world… We do not have enough schools. In every part of the Colony. where they could obtain pupils, free primary schools should be established, and parents should be persuaded or compelled to send their children to such schools.[3]

While the establishment of English schools was generally well received by the government, Chinese and other vernacular schools were left to the communities concerned. This pattern of partitioning the educational structure along communal lines proved to have severe economic and political consequences in the years to come. As the economy grew rapidly, it became clear that the general level of education did not keep pace with economic development of Singapore. The development of a modern education in British Malaya, including Singapore, was dictated by an education policy based on the goals and expected outcomes of colonial rule. The situation here was more complex than in other colonies like Egypt and the Gold Coast due to Britain's differential commitments to various ethnic communities.

EDUCATIONAL REFORMS AFTER 1870

When the control of the Straits Settlements was transferred to the Colonial Office in 1867, the new government felt the need to investigate the level of educational progress made in Singapore under the Indian Office. In 1870 Governor Sir Henry Ord set up the Woolley Committee, which reported that "the state of education in the colony has been and is in a backward state … arising in a great measure from the indifference of the different races, more particularly the Malays, to receive instruction, and to the want of sufficient encouragement from the Government itself".[4] Even on the training of pupils to become clerks in the government and mercantile sectors, the committee found that while "most of them are competent to work out a simple sum in arithmetic and to copy English in a good legible hand, ... as a rule they have no ideality: ideas they have none and they are quite incapable of expressing themselves in writing either grammatically or logically".[5] The committee recommended the appointment of a Director of Education to take charge of schools, the expansion of vernacular instruction, and the improvement of education for girls. However, it was the provision of English education which most clearly exposed the apathetic effort of the Straits Government.

Throughout most of the nineteenth century the Straits Government was not directly involved in promoting English education. It only subsidized English-medium schools, most of which were run by Christian missions. Widespread provision of English schooling was not favoured by the British colonial administrators because of the background of the indigenous society. The British Resident of Perak, Frank Swettenham commented:

> One danger to be guarded against is to teach English indiscriminately ... I do not think it is all advisable to attempt to give the children of an agricultural population an indifferent knowledge of a language that to all but the very few would only unfit them for the duties of life and make them discontented with anything like manual labour.[6]

Such prejudice was not uncommon amongst British officials serving in the tropics. In 1891 a Governor-General in India stated to the same effect that "original scientific research demands mental and physical qualifications which are not apparently found in races bred in tropical climates".[7] But it was obvious that the need for the English language was crucial in understanding the technical knowledge imported into Singapore by the Europeans.

A deciding factor affecting the success or failure of the government's educational policy since Raffles first commented in 1823 that "education must keep pace with commerce" was financial support. A meagre amount of the government's revenue was set aside for education in Federated Malay States and the Straits Settlements, as can be seen from Table 8.1. As an indication of the British commitment to educational progress, in the last quarter of the nineteenth century, the average expenditure on education in the Federated Malay States was about 1 per cent of the total revenue. As a result, teachers were underpaid and earning less than policemen or peons. In 1891 the average salary of a Malay teacher was only $11 a month, and

Table 8.1
Expenditure on Education in the Federated Malay States, 1875–1900

	Total Revenue ($)	Expenditure on Education ($)
1875	409,394	—
1896	8,434,083	—
1898	9,364,467	96,699
1899	13,486,410	106,588
1900	15,609,807	139,059

Source: K. Watson, "Rulers and Ruled: Racial Perceptions, Curriculum and Schooling in Colonial Malaya and Singapore", in *The Imperial Curriculum: Racial Images and Education in the British Colonial Experience*, edited by J.A. Mangan (London: Routledge, 1993), p. 159.

the government made no attempts to recommend any increase because "with the larger number of vernacular schools there would be increased competition for the posts, and competition would keep the salaries down".[8] Not surprisingly, the rate of resignation of European and local teachers from the vernacular and government schools was high.

The wealthy Straits-born Chinese in Singapore, Malacca and Penang pressed for greater exposure to English education. They came to be known as the "Queen's Chinese", stoutly pro-Queen Victoria, sent their sons to England to be educated and looked upon the Straits Settlements as a county of England. Writing in the *Straits Chinese Magazine* at the end of the nineteenth century, Dr Lim Boon Keng said:

> A good English education is, no doubt, the best legacy a Chinese or any other parent in the British Empire can leave to his children. But what is the sort of education the majority of our youths get? ... I regret to say that the "English education" is a very poor one so far as the majority of the boys are concerned.[9]

In the same article, Lim also criticized Chinese education in Singapore because it centred too much on the study of the Classics. He recommended "a more liberal course of studies, including science and mathematics" so that Chinese schools would "become the complement of the English schools".[10] The government's failure to recognize the need to improve English education coincided with the rapid expansion of modern Chinese schools whose outlook was focused towards China and restricted the mobility of pupils within the Chinese community of the Singapore's economy and society at large. Inspiration was obtained from the reforming movement of scholar-official, Kang Yu Wei in late nineteenth-century China.

The overall effect of the slow changes in the level of education in Singapore was the strengthening of divisions in society, separating the English and the vernacular education and accentuating racial, cultural and linguistic differences. In such an environment, it was difficult for Western technical knowledge and skills to diffuse successfully. At best, some English-educated boys were trained in middle-level jobs in the modern sector such as surveying, typography, draftsmanship, telegraphy, and machine repair. However, for the majority, acquiring the English language was the key to better opportunities in administration and commercial work. In short, the training of the human resource of Singapore at this time fulfilled the needs of the entrepôt economy rather than in creating a vital workforce necessary for large-scale industrialization.

A step in the right direction, however, was made by Governor Sir Cecil Clementi Smith when he introduced two measures which had great long-term significance for Singapore. These were the founding of two Queen's Scholarships in 1889 and the introduction of the Senior Cambridge Examination that still remains a critical component of Singapore's school examination today. The creation of the Queen's scholarships was a vital step to the rise of the Western-trained professional Chinese elite who were to exert a profound effect on the interaction between the Chinese society in Singapore and the Western administration in the twentieth century. In the years 1886 to 1888, five able young men were selected for the scholarship. Out of the five, two were Straits-born Chinese, namely Song Ong Siang and Lim Boon Keng. The other three recipients were J. Aitken who, on his return, started a legal firm with Song Ong Siang, C.S. Angus who studied Engineering, P.V. Locke of Penang who became an editor of the *Straits Chinese Magazine*, and D.A. Aeria.

Not everyone in the European community, however, appreciated Clementi's educational innovations. They were criticized by the *Singapore Free Press*, for concentrating resources at the top and "ignoring the broader base of primary education".[11] Clementi's strategy was in line with the so-called "Downward Filtration" theory that was strongly adhered to in British India during the latter half of the nineteenth century. The Court of Directors of the East India Company supported this theory:

> The improvements in education, however, which [have] most effectively contributed to elevate the moral and intellectual condition of the people, are those concern the education of the higher classes; of the persons possessing leisure and natural influence over the minds of their countrymen. By raising the standard of instruction amongst these classes, you would eventually produce a much greater and more beneficial change in the ideas and feelings of the community than you can hope to produce by acting directly on the more numerous class.[12]

At the start of the new century, there was no reason for the British to change their *laissez-faire* education policy so faithfully adhered to in the nineteenth century. Indeed, the imperial policies of France, Germany, and Britain on the issue of education in their colonies had certain similarities. All three powers were concerned with the economic exploitation of the colonies and the best utilization of their resources — including human resources. German colonists in Tanganyika, in the early years of the twentieth century, "needed office assistance who could read and write, and the presence, in important centres in the interior, of natives who understand German and could interpret government decrees to their fellow-countrymen".[13] For the British in Malaya and Singapore, the basic aim, as in the nineteenth century, was to strike a balance between the provision of sufficient English schooling to meet the colonial administrative and commercial manpower needs. Throughout the nineteenth century, the Singapore colonial administration was more than content to exploit the missionary societies and get education on the cheap. A few English mission schools were partly supported through a system of grants-in-aid by the government.

In 1902 the Kinnersley Committee was set up by the government to review education in Singapore. The level of English education was summed up in the report:

> All or very nearly all, of the European and Eurasian boys attended an English school, at any rate for a time. In the town a large proportion of the Chinese boys go to an English school and learn a little reading and writing, but very few go through the whole course. If all the boys mentioned attended English schools and remained for seven years, there should be 14,549 names in enrolment, whereas there were actually only 7,708 boys in enrolment at the English school last year.[14]

The committee also introduced an Education Code which dictated the nature of educational policy right up to the end of the First World War in 1919. The main thrust was on the provision of English-medium primary schools. Though there was no ethnic restriction, improvement in English education largely stagnated because of limited school facilities and enrolments. It also continued to be a prerogative of the locally-born boys of successful Chinese and Indian merchants. The cost of an English education was beyond the means of the indigenous masses who were mostly employed in menial occupations.

DEVELOPMENTS IN INDUSTRIAL AND HIGHER EDUCATION

The slow rate of progress in English education was matched by the miserable development in technical education. The demand and supply of technical education was closely linked to the overall educational policy of the colonial administration. It was also dictated by the manpower needs of the colonial government at that time. The expansion of British political and economic interests in the Malay Peninsula after 1874 and the subsequent explosion of the immigrant population in the Straits Settlements had increased the demand for lower administrative personnel in the government and commercial sectors. In the 1899 *Annual Report on Education* of the Straits Settlements, it was reiterated that:

> In a sense every English School in the Straits Settlements is essentially a commercial school in that it is mainly attended by boys who study the language for its commercial value and not as a mental training or road to higher work.... The criticism that our schools are too exclusively literary and trend too much towards clerical employment is one [which] ignores the important fact that the interests of the Straits Settlements are mainly commercial.[15]

The priority of the educational policy at the turn of the century was the training of a Westernized, English-speaking pool of clerks and government officers, so vital to an efficient running of Singapore's entrepôt economy. However, this is not to say that technical education was totally ignored.

Industrial education was encouraged in the 1880s by the grant of a four-year Industrial Scholarships for boys who wished to become engineers, engine-drivers, or surveyors. They were required to undergo apprenticeship in practical engineering with the Tanjong Pagar Dock Company and British engineering firms, such as Riley and Hargreaves or the English

Printing Office in Singapore. But the response to these scholarships was disappointing, as shown in the Return of Industrial Scholarships and Apprenticeships for the years 1888 to 1899, as shown in Table 8.2. They failed to attract the best candidates. A number of the candidates were disqualified and failed in the examinations. Consequently, the number of scholarships offered declined over the years and in 1893 and 1896 none were offered, presumably due to poor demand. Table 8.3 provides some information on the occupational background of the applicants, the majority of whom were engineers and draughtsmen. Many of the candidates were working on board ships or were employed in the municipality.

Though the response from private firms was dismal the few individuals involved were from industries which were well established, namely the printing presses and the British engineering firms of Riley and Hargreaves, and Howarth Erskine and Company. In 1899, a complementary overseas scheme was introduced to send engineering and surveying apprentices to the Roorkee Engineering College in the Saharanpur District of Uttar Pradesh, India. After two candidates were selected, the scheme was abolished "on the grounds of expense".[16]

In the twentieth century, the issue of technical education and vocational training gradually attracted the administration's attention. In

Table 8.2
Return of Industrial Scholarships and Apprenticeships

Year Given	No. of Scholarships	No. of Candidates	No. of Scholarships Offered
1888	13	10	9
1889	5	8	5
1890	5	5	5
1891	4	4	4
1892	5	5	5
1893	—	—	—
1894	9	7	6
1895	1	1	1
1896	—	—	—
1897	9	10	9
1898	1	3	1
1899	4	2	2

Source: Proceedings of the Legislative Council of the Straits Settlements, No. 19, 25 April 1899, C117.

Table 8.3
Occupations of Candidates Applying for Industrial Scholarships

Occupation	No.
Engineers on board ships	17
Draughtsmen from Municipality	3
Civil Engineers from Municipality	2
Bill Collector from Municipality	1
Compositors from Government Press Office	2
Compositor from printing firm	2
Apprentices from engineering firms	3
Tailor from Penang	1
Teacher from Siak	1

Source: Proceedings of the Legislative Council of the Straits Settlements, No. 19, 25 April 1899, C117.

1913, speaking on the educational needs of Singapore, the Reverend W. Runciman suggested that Singapore and the Straits Settlements needed a technical college or institute.[17] He proposed that the institute should provide for the teaching of engineering in its various branches and also for the teaching of mining and agricultural sciences. The official response was lukewarm and characteristic of the colonial administration at this time. The Colonial Secretary, R.J. Wilkinson, was certain that "if an elaborate engineering institute were started by the Government it would call for considerable sums of money for bursaries to encourage boys to attend".[18] Hence, consideration of high expenditure and the perception that technical education was not popular or necessary for the indigenous population were the key inhibiting factors. Nevertheless, a committee was set up in 1919 to look into the provision of technical and vocational education after some lobbying by the Straits Chinese British Association. Under the chairmanship of A.H. Lemon, the committee concluded that "the need for a technical school for training subordinates has become a matter of urgency, unless the [government] departments are to be dependent on obtaining subordinates who have technical training in other countries".[19] It also recommended the establishment of a trade school to allow boys recruited as artisans and mechanics in private firms to receive some form of formal training. The main problem, as pointed out in the Report, was the poor employment prospect of those who hoped to pursue or who had undergone formal technical training. Jobs were generally limited to the various government departments. Salary schemes for trained technicians were less attractive than those of a clerk who "has shorter hours, indoor work and other advantages, while the technical man ha[d] to work in the open often in the jungle or in some remote station".[20]

Little was done to implement the recommendations of the Lemon Committee. It took another six years before the British administration decided in 1925 to form another committee under Richard Winstedt to consider the question of industrial and technical education in Singapore. The Technical Education Committee considered the problems of technical and industrial education at four levels, namely higher technical education leading to the attainment of degrees, intermediate technical education, education in trade schools, and industrial education for children. The provision of higher technical education was unanimously rejected by the committee because, supposedly, the cost outweighed the benefits.[21] As stated in the Report, shown in Table 8.4, within the Straits Settlements, there were only 866 immigrant and locally-born males, or 0.43 per cent of a population of 197,588, in professions which required a higher technical education. Presumably, the majority of this group was not in possession of a degree of some sort since the only access to higher education at this time was through the Queen's Scholarships. The same argument was used that it was not cost-effective for the government to provide higher technical education which affected only a small percentage of the population. To support its stand, the committee highlighted the failure of a similar attempt in Colombo where a Government Technical College was set up as early as in 1893 to teach civil, mechanical, and electrical engineering. Due to the lack of students, in 1910, the college was converted to a technical school for apprenticeship training of subordinates in government departments.

Table 8.4
Number of Professionals in the Straits Settlements in 1925

Profession	No.
Consulting Engineers	3
Architects	78
Civil Engineers and Surveyors	425
Mechanical Engineers	360

Source: "The Winstedt Report 1925" in Wong and Gwee, *Official Reports on Education*, p. 93.

Politics of Imperial Education

Surveys were carried out by the Winstedt Committee and the findings revealed the poor aptitude and response of the population towards the acquisition of technical skills and the learning of science-based subjects. Evening classes in mathematics, physics, chemistry and mechanics were started at Raffles Institution in 1922 and the attendance for the period 1922–23 was most discouraging, as shown in Table 8.5. The Report commented that employers and employees alike were not giving the evening classes their full support. Similarly, the concept of trade schools received lukewarm responses from the public, as shown by the following comments:

- Trade schools are premature. What trades have we? There are the trades of the Harbour Board, the United Engineers and the Central Engine Works.
- It is doubtful if there would be a demand for superior craftsmen with a Trade School training or if Eurasian would enter. There are no prospects yet for a Trade School.
- I [a respondent] am not sure if pupils would attend.

Although Singapore was regarded as "the best place" for the establishment of more than one trade school because its male population between the ages of 10 and 14 numbered between 12,000 and 13,000 at the census of 1921, the Committee concluded that "many of the youth of this Colony have a genuine distaste for hard and continuous manual labour" and, moreover, "they have no desire for technical training".

The road towards the provision of higher education by the Colonial Government was likewise filled with obstacles and uncertainties. Hitherto, the only channel for higher education was through the Queen's

Table 8.5
Attendance in Science-based Subjects, 1922–23

Subject	Attendance
Science (Theoretical)	10 students
Science (Practical)	15 students
Chemistry and Physics	10 students
Mathematics and Mechanics	too few to form a class

Source: "The Winstedt Report 1925" in Wong and Gwee, *Official Reports on Education*, p. 95.

Scholarships. Between 1889 and 1911, forty-five students became Queen's Scholars, a rare distinction at that time. But the scholarship soon faced strong criticism from the government. From 1908 onwards, instead of two scholarships, only one was offered because "it [was] intended to expend the money saved by the abolition of one scholarship on the improvement of education in other directions".[22] After 1911, the Queen's Scholarships were abolished altogether as the government felt that it was not justifiable to provide an overseas education for a handful of boys at the expense of the majority. But behind the official explanation was the British fear that "the aspirations of the sons of the soil were likely to create a situation such as exists in India and other progressive countries, where the permanent population is demanding a large share in the administration of the country of their birth through their educated members".[23] To the ruling British, higher education and English education in general possessed an aura of power and rarity; only a minority and privileged few could enjoy it. The Queen's scholarships were not resumed until 1923. And the purported "improvement of education in other directions" was not carried out. The late Victorian drive towards creating "developed estates" in the periphery of the Empire was waning. No further change and expansion was thought necessary and ideas for improvement trickled off.

The only noteworthy development was the decision by the government to establish a college for higher education, to be known as Raffles College of Arts and Sciences, to mark Singapore's centenary. Richard Winstedt, the Director of Education, was appointed as Principal in 1921, while retaining his former post. In carrying out his task, Winstedt claimed that "Raffles College took more of my thought and time, and I hoped I set its infant footstep on the right way".[24] Unfortunately, its construction was delayed and when it was finally opened in 1928, the College operated as a tertiary institution for teacher training at diploma level. According to Tregonning, the Raffles College "failed utterly to satisfy the aspirations for higher education held by so many of the community".[25] A contemporary observer maintained that the College's unsatisfactory progress was due to "the failure of the Government to discharge its very obvious duty of appointing a Principal even before the buildings were completed or at least immediately after".[26]

After nearly a decade of doldrums, the Legislative Council of the Straits Settlements decided in 1938 to look into the prospect of forming an engineering department within the Raffles College. This awakening was due to the fact that "there ha[d] been in recent years in most countries a great increase in the number of engineers employed in communications, owing to developments in wireless and automatic telephony, etc.".[27] But the obstacles, as stated in the Report, were serious:

> The first is that local boys who have obtained the Junior or Senior School Certificate prefer to take up clerical appointments and disdain manual labour. The second is that there is no employment in industry available for the boys of those standard of education ... It is true that vocational or technical education is in itself not a means of creating employment: on the other hand, if the facilities for it are available, it may provide an encouragement to well educated boys to enter industry in the expectation that by acquiring technical qualifications they may make some progress in their trade or profession.[28]

Another disadvantage pointed out by the Council was the absence of an effective apprenticeship system in Singapore and Malaya because "there is no electrical engineering industry and a limited amount of mechanical engineering" and a lack of "competent supervisors, foremen, and clerks of works, etc. as exists throughout Britain".[29] Nevertheless, the issue of providing engineering education was serious enough to warrant the setting up of a committee under William McLean in 1939 to look into the progress of Raffles College and the possibility of raising its status to a fully-fledged university. The McLean Committee reported:

> There is no doubt in our opinion that the College has made progress since its inception; equally there is no doubt that this progress must proceed further before the College can fulfil the ideas of its founders that it should become the nucleus of a University. Clearly at the present time neither the standard of achievement of its students, nor its contributions to knowledge fulfil the conditions which we have laid down.[30]

Specifically on the question of establishing a school of engineering as a faculty of Raffles College, the committee concluded that "at the present time a Faculty of Engineering at Raffles College should not be established" because in a "non-industrial country" such as the Straits Settlements, the demand for mechanical, electrical and civil engineers is "insufficient to justify the large capital and current expenditure".

The only bright spot, as highlighted in the McLean Report, in Singapore's otherwise uneventful educational development was the progress made by the King Edward College of Medicine. As was often the case, its formation in 1905 was the result of concerted pressure by the Straits Chinese and their generous donations. In 1912 the school was named the Edward VII Medical College. To match its new status, Dr J. Argyll Campbell, a Vans Dunlop scholar in pathology and surgery, was recruited from England, being appointed Professor of Physiology in 1913.[31] In 1920 the College was further upgraded to become King Edward Medical College with a full-time teaching faculty. Through the years the College built up its reputation and "its diploma in Medicine differ[ed] not greatly in its standard from that of those awarded by Licensing Bodies in Great Britain, while the aggregate contributions of the members of its staff to knowledge by research command[ed] respect".[32]

In summary, it is not uncommon for colonial administrations in British tropical colonies to largely neglect the provision of Western education except to the extent that it was necessary to fill the lower clerical and administrative grades of the government services and the private sector. Although considered by the London parliamentarians as an important British trading port-city in the Far East, the indigenous population did not benefit much from the transfer and diffusion of Western knowledge. While it was not entirely due to the government's ineptitude, the impact of this policy on manpower development and subsequent attempts at industrialization was serious. What factors directly or indirectly blocked the advancement of education in the colony?

OBSTACLES TO EDUCATION CHANGE: MEN-ON-THE-SPOT

The existence of a multi-ethnic and multi-cultural plural society could be seen as a powerful barrier. As the majority, the Chinese set up schools to promote Chinese education along traditional Confucian lines. Denied government finance, the Chinese proceeded to build up their own vernacular primary and secondary school systems modelled on those in China, often using textbooks recommended by the Chinese nationalist government. Culturally, Chinese nationalism, especially during the early decades of the twentieth century, created a renewed consciousness of Chinese identity. Hence, fuelled by a political stance towards foreign domination, the Chinese generally adopted an uninspiring attitude towards English schooling. The Indians were also left to establish their own Tamil-medium schools which were oriented towards political developments in India. The educational standard of these schools was elementary and many of the pupils ended up as unskilled labourers.[33] Sir George Maxwell, the Chief Secretary of the Federated Malay States, for example, wrote in the 1920 *Annual Report of the Straits Settlements*, regarding the objectives of educating the Malays in government schools:

> [T]he aim of the Government is not to turn out a few well educated youths, nor a number of less well educated boys: rather it is to improve the bulk of the people and to make the son of the fisherman or peasant a more intelligent fisherman or peasant than his father had been, as a man whose education will enable him to understand how his lot in life fits in with the scheme of life around him.[34]

However, the majority of Malays were not keen to send their children to government schools, preferring the Koranic schools run at the village level. While vernacular education of the various races acted as barriers to the acceptance or acquisition of English education, the perceptions and inconsistencies of British colonial administrators were largely responsible for the stunted growth of education in colonial Singapore.

The introduction of modern education in British Malaya, including Singapore, was essentially a by-product of colonial rule. It was characterized with the general assumption that, in carrying out the "white man's burden", Britain was bestowing on uncivilized inhabitants of her colonies the benefits of a paternalistic motherland. These perceptions of European superiority were certainly shown by administrators controlling the peripheral lands of the British Empire. The Malays and the thousands of unskilled Chinese and Indian labourers who made up the working class population were considered unsuitable for Western education. Addressing a group of former colonial officials at the Royal Colonial Institute in 1896, Frank Swettenham, one of the longest serving officials in British Malaya, provided an indication of such an attitude:

> A native of the East is curiously prone to imitate the Western, but his imitation is nearly always only partial... He clothes himself in items of the European dress, he learns scraps of the language, essays British sports, without sufficient energy or determination to thoroughly succeed, and he will even, with what seems praiseworthy enterprise, take up the planting of some new product in imitation of an European neighbour, often, I regret to say, wasting thereby a capital that would have been better

employed in some other form of planting or business which he really understood... I think it should be our object to maintain or revive his interests in the best of his traditions, rather than encourage him to assume habits of life that are not really suited to his character, constitution, climate, or the circumstances in which he lives - which are, in fact, unnatural to him, and will lead him to trouble and disappointment, if not to absolute disaster.[35]

In the eyes of the British official, vernacular education was encouraged in order to develop character and citizen training. However, except for Malay education, education of the Chinese and Indian population remained outside the responsibility of the colonial government. But when Chinese and Indian education created a sense of ethnic awakening in the 1930s, the British administration quickly imposed restrictions on the nature of the education and limited the flow of immigrants.

Swettenham's comments reflected the westerners' cultural perceptions of the capability of indigenous peoples in Asia and Africa to understand the literacy world of the industrialized West. From the late eighteenth century on, the European "articulate classes", notably the British and the French colonial administrators, natural scientists, social theorists, and anthropologists embroiled themselves in admiring their cultural superiority and in formulating attitudes and ideologies relating to the transfer and diffusion of science and technology to non-Western world.[36] African peoples were seen to have no mental faculty to understand scientific thinking, whereas most European observers believed that the Orientals, mainly the Indians and the Chinese, were capable of absorbing the technology and scientific ideas of the West. Thus, in British India, for example, English education was introduced at all levels and became the means of unlocking the secrets of the Western sciences to the locals.

As one of the longest-serving colonial administrators, Swettenham commented that "we should not aim at giving Malays the sort of higher education that is offered by the Government of India to its native subjects, but I would prefer to see the establishment of classes where useful trades would be taught".[37] He was reinforcing a perception held by the colonial administration since the early decades of the nineteenth century:

> It must be stated, that of all Asiatics, the Malay is probably the least susceptible of improvement from Education. Those engaged in the heavy task of opening the mind of a genuine Malay are almost in despair at finding so little mind to work upon. With an infusion of Indian or Chinese blood, the case is different, but the pure Malayan intellect is

very stagnant, and what is worse, the inferiority is felt and acknowledged by themselves.[38]

Another administrator, Frederick Weld, who served as Governor of the Straits Settlements from 1880 to 1887, made it clear that "the common Chinese feeling is that we — an eccentric race — were created to govern and look after them, as a groom looks after a horse, whilst they were created to get rich and enjoy the good things of the earth".[39] British administrators in the Straits Settlements believed that the British were the better judges of the interests of the native population. Such a perception was famously echoed by the imperialist Lord Milner in the House of Lords on 12 May 1920 - that the British "are better judges, for the time being, of the native population than they are themselves".[40]

As the Assistant Director of Education in the Straits Settlements and Federated Malay States from 1916 to 1923 and Director of Education from 1924 to 1931, Richard Winstedt's long period in the field of education failed to produce anything of lasting value. On the contrary, his actions impeded developments. Two clear instances may be cited. First was his encounter with the American-backed Methodist Mission. By 1915, the Methodist Mission had set up seven schools offering English and vernacular education in Singapore. From 1914 to 1920, the Mission was planning to establish a college and negotiations were started with British local officials and in London. The possibility of having a new college was enthusiastically received by the Straits Chinese and funds were easily raised; a principal designate was also named. The Colonial Office approved the plan, but it was blocked by Winstedt. Fearing the spread of American influence and its erosive effect it might have on British prestige in the Malay Peninsula, Winstedt was also able to convince the Governor, Laurence Guillemard, to support his stand.[41] This episode reflected the fear, jealousy or inadequacy of a British official dealing with the growing competitive power of the United States. By the end of the nineteenth century, Britain no longer led the world in technology; the United States and Germany had both overtaken her.[42] Of course, one outcome of Winstedt's sabotage was the need to pacify the local community's desire for some form of higher education. Hence, the decision was taken to set up Raffles College in 1921. But Winstedt was made the principal, and, as mentioned above, his appointment delayed the development of the College.

The second situation concerned Winstedt's formulation of education policy for the Malay masses. As Assistant Director of Education in 1916, Winstedt made an education study tour of other imperial colonies, namely, the Philippines and Java. Returning in 1917, he proudly proclaimed that a new education policy which he called "the new education" would be implemented. For someone who claimed to be knowledgeable in Malay culture and literature, what Winstedt recommended "can only be interpreted as a blatant policy to stifle Malay intellectual development and political consciousness, thereby safeguarding the vested interests of British imperialism".[43] His observations convinced him that Malay education in Malaya and Singapore had been too advanced as compared to the colonial education in the Philippines and Java. He agreed with the Dutch that history should be thrown out of the curriculum and that the basic feature of Malay education should be "rural" with a "strong manual, agricultural bias". Winstedt's retrogressive policy remained influential in Malaya until the Second World War.

The deliberately limited provision of English-stream education and, until the 1930s, a liberal policy towards the spread of vernacular education gave rise to a cleavage between a privileged Westernized, English-speaking local elite and the masses at large. In most cases, only the sons of Malay chiefs, rich Chinese and Indian entrepreneurs had the opportunity to enjoy English education which paved the way to careers in lower- and middle-level administrative posts in the government services. Hence, a dualistic social structure became evident especially amongst the Chinese. The Straits Chinese gradually assumed part of the leadership of the Chinese community. In the words of the British official, W.G. Shellabear, they were "the most highly educated and the most influential section of the Chinese community in the British possessions".[44] However, their social position was shared by a handful of illiterate or half-literate Chinese migrants who amassed wealth through sheer hard work and luck. As a show of gratitude to society, many of these *towkays* (rich Chinese businessmen) contributed generously to promote Chinese education and charitable work.[45] With the prestige and power they obtained, the Western-educated and the rich Chinese differentiated themselves from the masses through their extravagant lifestyles.[46] For the thousands of unskilled and illiterate immigrants, education was not the key to fortune and good life; a strong will, the ability to communicate in languages and dialects commonly used, an adventurous spirit and lots of luck were looked upon as essential ingredients of success.

To be fair to these colonial administrators, educational policies in Singapore before 1940 "fluctuated" according to changing practices and trends in England. Inevitably, British officials posted to Singapore and

British Malaya adhered to the prevailing policies and failed to use their own creativity in initiating far-reaching changes. The 1902 Education Act in Britain, for example, led to an expansion of grammar secondary schools specifically aimed to produce the manpower for white-collar jobs and not jobs related to industry. Colonial officials posted to the periphery of the British Empire were influenced in one way or another by such developments in England. This was especially obvious in the sphere of technical education and university education.

In Britain, before engineering education gained importance towards the end of the nineteenth century, the acquisition of technical skills was gained through a period "of social formalisation and apprenticeship, in which a pupil was indentured to a master for a period of years".[47] The forces of rapid industrialization, however, resulted in the development of a system of technological and scientific education in schools and universities in most of the European countries, particularly in France and Germany, and also the United States. Since the mid-eighteenth century, the early French schools had already developed operative methods such as "industrial science" and practice-oriented teaching programmes were introduced in reputable institutions such as Ecole des Mines, Ecole Centrale des Arts and Ecole Polytechnique.[48] To meet the increasing industrial prowess of Europe and the United States, from the 1870s, Britain created new institutions and courses to cater for technical and science education and, at the same time, continued its apprenticeship system. Mechanical institutes and scientific and technical schools were established at universities. However, the development of technical education in Britain was strewn with problems and "a few more decades were to pass before a scientific-technical and scientific-industrial system of education developed" by the 1880s.[49]

Unlike France, Germany and the United States, engineering studies at the university level only appeared in Oxford or Cambridge in the 1870s. The academic preference for staff and students alike, however, was still in the non-science courses. Engineering education was looked upon with disdain because it was associated with the artisan class. Some industrialists viewed applied science and technical education with scepticism and clung on to the traditional apprenticeship system. Technical schools in their early decades were seen more as "places of correction than as centres of skilled instruction".[50] Such attitudes were found, for example, in the British shipbuilding industry before 1914:

> [T]echnical education for skilled workmen was deemed beneficial because it discouraged vandalism, promoted moral strength, and broadened a man's outlook as well as giving him a better grasp of the job. It was important not so much because it imparted a better knowledge of the principles of shipbuilding and engineering, but because it helped to inculcate habits of good conduct.[51]

The intended outcome of technical education for the workmen in the British dockyards echoed the general view of the colonial officials in British Malaya mentioned above, namely that education for the people was aimed basically at character training. The frequent remarks made by the officials that employers in Singapore were not eager to send their workers or even themselves to some classes in technical instructions were also encountered in Britain, as reported by the Association of Teachers in Technical Institutes in 1927: "Manufacturers were suspicious or indifferent. They feared their trade secrets would become known, or they saw no immediate benefits, or they had no faith that schools or colleges could be of service to productive industry".[52]

In summary, the upgrading of English and technical education in Singapore during the years 1900 to 1940 was hampered by educational policies which were essentially parsimonious in nature, and implemented by administrators who harboured the perception that the spread of English education must be controlled. This was more conspicuous after the First World War when British officials who served in the colony handled matters more "by reason of their political fear and anxiety and their professional selfishness and obscurantism" than a genuine interest in the growth and development of Singapore.[53] In spite of increasing annual expenditure on education, by 1937 only 12 per cent of the children of suitable age attended English schools and, in line with the Home Government's ideal that public expenditure in dependent colonies should be kept as low as possible, the Colonial administration exclaimed that "the 5.69 per cent of the total revenue of the Straits which was spent in 1931 on education was a smaller percentage than in other important colonies and dependencies in the Empire".[54] Such comments seemed to reinforce the concept of a colonial "nightwatchman" — the main function of colonial administrators was to collect revenues, balance the budget, and maintain law and order. Expenditure beyond what was necessary for performing the nightwatchman's tasks was discouraged. Thus, in the field of education in Singapore, up to the outbreak of the

Second World War and out of the population of school age children of about 200,000, hardly more than one-third were in schools and only one-third of this figure were in the English language stream.[55]

Notes

1. G.G. Hough, "Notes on the educational policy of Sir Stamford Raffles", *Journal of the Malaya Branch Royal Asiatic Society* XLII (1969): 174.
2. M. Rudner, "Colonial Education Policy", in *Development Studies and Colonial Policy*, edited by B. Ingham and C. Simmons (London: Frank Cass, 1987), p. 194.
3. *Straits Times*, 4 December 1867.
4. *Report of the Select Committee to inquire into the State of Education in the Colony*, 1870, C93.
5. Ibid.
6. Quoted in F. Wong and Y.H. Gwee, *Official Reports on Education: Straits Settlements and the Federated Malay States 1870–1939* (Singapore: Pan Pacific Books), p. 196.
7. S. Ambirajan, "Steam Intellect and the Raj: South India in the Nineteenth Century", in *The Steam Intellect Societies: Essays on Culture, Education and Industry, circa 1820–1914*, edited by Ian Inkster (University of Nottingham: Department of Adult Education, 1985), pp. 160–80.
8. C. Bazell, "Education in Singapore", in *One Hundred Years of Singapore*, edited by Makepeace et al., p. 463.
9. Quoted in Moore and Moore, *The First 150 Years*, p. 473.
10. Quoted in ibid., p. 474.
11. Turnbull, *History of Singapore*, p. 116.
12. Ambirajan, "Steam Intellect and the Raj", p. 164.
13. J.A. Mangan, "Ethics and Ethnocentricity: Imperial Education in British Tropical Africa", in *Sport in Africa: Essays in Social History*, edited by W.J. Baker and J.A. Mangan (New York: Africana Publication House, 1987), p. 157.
14. *Report of the Commission of inquiry into the System of English Education in the Colony*, 1902, C99.
15. Quoted in D.D. Chelliah, *A History of the Educational Policy of the Straits Settlements with Recommendations for a New System based on Vernaculars* (Kuala Lumpur: Government Press, 1947; reprinted Singapore: Kiat & Co., 1960), p. 105.
16. Bazell, "Education in Singapore", p. 474.
17. *Straits Times*, 28 July 1913.
18. Ibid.

19. Quoted in Wong and Gwee, *Official Reports on Education*, p. 73.
20. Ibid., p. 92.
21. *Report of Technical Committee 1925*, 1926, C11.
22. Bazell, "Education in Singapore", p. 471.
23. A.H. Carlos, "The Eurasians of Singapore", in *One Hundred Years of Singapore*, edited by Makepeace et al., Vol. 1, p. 369.
24. R.O. Winstedt, "The Land I Love", *British Malaya*, September 1940, p. 71.
25. K.G. Tregonning, "Tertiary Education in Malaya: Policy and Practice 1905–1962", *Journal of the Malayan Branch of the Royal Asiatic Society* 63 (1990): 2. In comparison, universities and specialized institutions were established by other European imperialists in Southeast Asia. The French started a small university in Hanoi in 1917, the American established in 1908 the University of Philippines, the Dutch formed scientific and technical institutions specializing in medicine, law and engineering, and in Siam, in line with the government modernization programme, the Chulalongkorn University was set up in 1917.
26. C.E. Wurtzburg, "Higher Education in Malaya", *British Malaya*, February 1940, p. 208.
27. *Report on Education in Engineering by Professor G. McOwan*, 1938, C210.
28. Ibid., p. 216.
29. *Report on Education in Engineering*, 29 August 1938, C215 and C217.
30. Quoted in Wong and Gwee, *Official Reports on Education*, 1980, p. 143.
31. *The Times*, 15 May 1913.
32. Quoted in Wong and Gwee, *Official Reports on Education*, 1980, p. 143.
33. Turnbull, *A History of Singapore*, p. 118.
34. *Annual Report of the Straits Settlements, 1920*.
35. Quoted in Kratoska, *Honourable Intentions*, p. 186.
36. G.M. Young, *Victorian England: Portrait of an Age* (Oxford: Oxford University Press, 1964).
37. Quoted in Kratoska, *Honourable Intentions*, p. 186.
38. *Annual Reports of the Straits Settlements*, 1856–57, p. 14.
39. Quoted in Kratoska, *Honourable Intentions*, p. 186 p. 46.
40. Quoted in Rupert Emerson, *Malaysia: A Study in Direct and Indirect Rule* (New York: Macmillan, 1937; reprinted, Kuala Lumpur: University of Malaya Press, 1964), p. 290.
41. Edwin Lee, "The Colonial Legacy", in *Management of Success: The Moulding of Modern Singapore*, edited by K.S. Sandhu and Paul Wheatley (Singapore: Institute of Southeast Asian Studies, 1989), p. 26.
42. D.S. Landes, *The Unbound Prometheus: Technological Change and Industrial Development in Western Europe from 1750 to Present* (Cambridge: Cambridge

Politics of Imperial Education

University Press, 1969), pp. 339–45. According to Landes, progress in higher and technical education in Britain stagnated by the early decades of the twentieth century. The quality of scientific and technical instruction also suffered and soon lagged behind Germany and the United States where a constant supply of well-trained personnel met the rising industrial demands for skilled workers. In his view, education should provide four kinds of knowledge: (1) the ability to read, write and calculate; (2) the working skills of the craftsman and mechanic; (3) the engineer's combination of scientific principle and applied training; and (4) high-level scientific knowledge, theoretical and applied. In all four areas, Germany represented the best that Europe had to offer; in all four, with the possible exception of the second due to its apprenticeship system, Britain fell behind.

43. S. Maaruf, *Malay Ideas on Development: From Feudal Lord to Capitalist* (Singapore: Times Books, 1988), p. 56. Progressive education for the Malays was also hampered by the attitudes of the Malay ruling elite who criticized those Malay youths who migrated to the towns and took up non-agricultural occupations merely because they had an elementary knowledge of the English language. Hence, feudal outlook and colonial ideology complemented each other, creating detrimental results for the Malay population in general.

44. W.G. Shellabear, "Baba Malay: An Introduction to the Language of the Straits-born Chinese", *Journal of the Royal Asiatic Society Straits Branch* 65 (1913): 52.

45. In the Hokien dialect, the term "*towkay*" means "rich Chinese men".

46. Maurice Freedman, "Immigrants and Associations: Chinese in Nineteenth Century Singapore", in *Comparative Studies in Society and History* III (1960–61): 28; Yen Ching Hwang, "Class Structure and Social Mobility in the Chinese Community in Singapore and Malaya 1800–1911", University of Adelaide: Department of History, Working Paper No. 15, 1983, pp. 9–11.

47. M.L. Guillou, "Technical Education 1850–1914", in *Where Did We Go Wrong? Industry Education and Economy of Victorian Britain*, edited by G. Roderick and M. Stephens (London: The Falmer Press, 1981), p. 181.

48. Ulrich Pfammatter, *Making of the Modern Architect and Engineer* (Berlin: Birkhauser, 2000), p. 295.

49. Ibid., p. 299.

50. D.H. Aldcroft, "The Economy, Management and Foreign Competition", in *Where Did We Go Wrong*, edited by Roderick and Stephens, p. 24.

51. P.L. Robertson, "Technical Education in the British Shipbuilding and Marine Engineering Industries 1863–1914", *Economic History Review* 27 (1974): 227.

52. Quoted in Guillou, "Technical Education", p. 173.
53. Lee, "The Colonial Legacy", p. 26.
54. Emerson, *Malaysia*, pp. 302–303.
55. T.R. Doraisamy, ed., *150 Years of Education in Singapore* (Singapore: TTC Publication, 1969), p. 38.

9
Technology Transfer and Limited Industrial Growth

By 1914, Britain ceased to be the only industrialized nation. The era of the "imperialism of free trade" had ended. The British government was increasingly concerned with the need to review and deploy the Empire's "underdeveloped estates", in Joseph Chamberlain's phrase, to support its domestic economy. Dependent colonies throughout the British Empire, however, were not encouraged to produce goods which could compete against imports coming directly from the industrial heartlands of Britain.[1] As suggested by Kemp and Headrick, so long foreign rule continues, colonial societies, including those that were developed solely as trading city-ports (such as Singapore), have generally failed to generate a favourable environment for industrialization to take place.[2] And even if some form of industrialization does take place, it is usually regarded as "peripheral" that is, industries that are largely dependent on imported technologies and supported by an abundant supply of cheap labour force, most of whom are women and children. Technology transfers which might have led to the growth of import-substitution industries were generally discouraged because not only they would have competed with British manufacturers, but they motivated the rise of native industrialists and cadres of engineers and technicians. Technology transfer, either in the form of a new process or piece of machinery, was introduced into a colony together with European experts who would zealously guard their knowledge and skills.

As Britain's major trading emporium in Southeast Asia, Singapore "was in no way an industrial city [and] at the end of the 1930s, first-stage import-substitution, involving a replacement of non-durable consumer goods, remained incomplete".[3] Industrial growth was linked to primary commodity exports and the rubber industry was the key contributor. Its role in the international trading system was that of a transshipment centre, just like Hong Kong which was used by British companies as the springboard to exploit the promising China market after 1842. Strategically situated, "[t]he colonial economy of Singapore and Malaya developed as part of what has been called the Old International Division of Labour (OIDL) in which the countries of the southern hemisphere exported primary products to the industrializing countries of the northern hemisphere which requited these with a return flow principally of manufactured goods".[4] Domestic industrial development in Singapore before 1940 (and certainly, in the 1950s too) was limited to production of goods which were closely linked to the rubber industry, tin-mining, shipbuilding and repairing industry. If one of the justifications for nineteenth century imperialism and twentieth century modernization effort was to spread western technology to the rest of the world and to promote "progress", why did the colony of Singapore fail to develop an industrial economy based on European technology and instead become what Headrick describes as "modern underdeveloped"?[5]

LIMITED INDUSTRIALIZATION

While the British and other European investors controlled the managing agency houses and set up all the major tin and rubber enterprises, the local (mainly Chinese) community also went beyond handicraft and cottage industries and started small manufacturing activities. By the early 1930s, the industrial undertaking in Singapore was an agglomeration of small Chinese firms in rubber milling, manufacture of rubber-soled shoes and boots, tyres, hoses, belting and other technical rubber goods, manufacture of sweets, medicines, biscuits, bricks, soap, flooring tiles, steel and rattan furniture, wire nails, pottery, coal tar and printing firms.[6] Table 9.1 shows the larger Chinese establishments which were successful enough to advertise their business operations in the 1928 edition of the *Singapore and Malaya Directory*. The majority of other races and Chinese firms were classified as general merchants, insurance agents, brokers, commission agents, printers, stationers, tailors, and timber merchants. This mixture of business establishments remained more or less the same up to 1940. Many of the Chinese entrepreneurs involved in manufacturing activities "relied largely on their ability as entrepreneurs rather than on any managerial or technical experience".[7] The employment of European engineers provided

Table 9.1
Chinese Manufacturing Activity in 1928

Type of Activity	No. of Establishment/Name
Cement Manufacturing	1 (Ho Hong Portland Cement)
Oil Manufacturing	1 (Ho Hong Oil Mills)
Soap Manufacturing	1 (Ho Hong Oil Mills)
Civil Engineering	2 (Lim & Seah; Wong & Co)
Motor Engineering	2 (United Motor; Eastern Auto)
General Engineering	1 (Tan Kah Kee & Co)

Source: *Singapore and Malaya Directory*, 1928.

the crucial technical expertise — as evident by the advertisements put up by small Chinese firms in the *Directories* of their presence. The engineering industry was dominated by the British operations of the United Engineers and the Singapore Harbour Board and a few smaller foundries, all of which had considerable number of European staff.[8] The rubber industry and the port development contributed significantly to the growth of the engineering sector. The demand for engineering and mechanical services to build new plants and manufacture products like sheeting machines, rubber mangles and scrap washers demanded by the thousands of rubber estates and numerous Chinese mills treating smallholder rubber led to the expansion of United Engineers.

From a demand perspective, Western firms did transfer modern machinery and techniques into the tin-mining and rubber industry mainly owned and operated by Europeans, as world demand for these products motivated them to diversify their economic interest into the production of tin ores and rubber. Former employees of agency firms, James Sword and Herman Muhlinghaus, incorporated the Straits Trading Company on 8 November 1887 and tin smelting operations were developed in Pulau Brani, an island opposite the New Harbour. British technical experts, such as T.E. Earle, John McKillop, John Carroll and S.B. Archdeacon, were the technical pioneers and introduced new smelting technology which made Singapore a world leader in tin ore production — the famous "Straits Tin" — by 1900. By 1914, the original twelve Cornish type coal firing furnaces were replaced by fifteen larger regenerative gas-fired reverberatory furnaces and, after 1928, all the furnaces were fired by oil.[9] With the rise in the value of tin beginning in 1898, European investors began to make their presence felt. With huge capital and modern hydraulic and dredging technology,

European companies soon dominated the industry, hitherto monopolized by Chinese miners. In 1912 the first tin-dredge was brought into operation and by the 1930s the tin industry in Malaya was Western dominated.[10] It is interesting to note that although the Europeans had access to imported mining machinery, early attempts to use them in mining in Malaya ended in failure. It was only when working on Chinese mines and adapting and developing mining machinery that European-owned mines could gain a good measure of success. However, the Chinese miners were also quick to adopt the modern mining machinery, although even under the best of circumstances, importing technology was a hazardous decision. By the early years of the twentieth century, contemporary observers noted that:

> The Chinese have not been slow to follow the example set them by their Western neighbours; and now no mine is regarded as properly equipped unless rails, trucks, and hauling engines are used to replace the coolie. Puddlers of various kinds are employed to disintegrate the *karang* [pay dirt] on its reaching the surface, and the old-fashioned coffin-shaped wash-box has given way to long sluice-boxes paved with riffles.[11]

The Rubber Estate Agency, incorporated in Singapore in 1906, was also instrumental in diffusing modern technology and latest methods, both in planting and factory processes, to no less than thirty-four plantation companies.[12] The utilization of updated technology was all the more important when the Singapore Chamber of Commerce introduced standard qualities comparable to London standards to facilitate the production and sale of Malayan rubber. Rubber milling factories in European-owned estates and Chinese smallholders in Malaya benefited from the development and manufacturing of a range of machinery, such as crepe and sheet rubber machinery, by Western engineering firms in Singapore that could adapted to the production needs.[13] For many Chinese smallholders, the need to upgrade their rubber sheets to standards required by European buyers resulted in the rise of rubber mangles manufacturing in Singapore. This activity was largely dominated by small Chinese firms in the 1930s. Finally, the upgrading of telegraphic and telephone network in cities and bigger towns in British Malaya, such as in Penang and Kuala Lumpur, during the inter-war years had the impact of making British Malaya into a single, efficient market for rubber. The communications network allowed agency houses in Singapore to keep in touch with their dealers and buyers up-country, including Chinese dealers.

Why did the British not look upon Singapore as a colony worthy of large-scale industrial development, particularly manufacturing for export? Like Hong Kong, the island had one of the best developed ports in the East and supported by a well-developed commercial sector. Moreover, one key decisions of the Imperial Economic Conference which met at Ottawa in August 1932 was to accelerate the industrialization process in Britain's dominions.[14] The conference was called to find ways of combating the worldwide economic depression by stimulating trade between countries of the Empire. The official answer was provided by a commission set up to investigate the state of Singapore's trade in 1933–34.[15] A strong inhibitive factor was the small size of the Malayan domestic market. The total population of about 4.5 million "may be large enough to support small scale industries, is not of sufficient size to enable a manufacturer to bring down his overhead charges by manufacturing on a large and economic scale, and a factory of any size must look to foreign markets to absorb the bulk of its products". The world depression which began in 1929 accelerated the imposition of protective tariff by other countries. It also convinced British politicians that the liberal principles of free trade which had been tenaciously followed for the past seventy years had to be abolished. A tax of 10 per cent *ad valorem* on all imports was imposed in 1931. The Report also commented that local manufacturers were dependent on Shipping Conferences which control ocean freights. Finally, it mentioned that the labour cost and cost of living was much higher than in Hong Kong because "labour generally speaking is imported and not indigenous and the labourer naturally expects to be paid more than he would be content with in his own [and] the cost of living is high owing to the fact that most foodstuffs have to be imported from overseas".[16] It cited that the main reason for the high cost of labour is the "high currency unit (which) has been regularly higher than the yen or the Hong Kong dollar and until lately than the guilder".[17]

In view of these adverse factors, the government concluded that: "The prosperity of Singapore has been built up on its entrepot trade. Industrial development is a later growth and has not begun to approach the entrepot trade in importance. To disturb this merely for the sake of protecting possible but still problematical industries would be to throw away the substance and grasp the shadow".[18] It was pointed out by the Commission that the measures to protect local manufacturers "is *prima facie* incompatible with the maintenance of the entrepot trade and it is difficult to see how the two can be reconciled".[19] The government's decision not to create a large manufacturing base as an integral component of the Singapore's economy was made known a few years earlier when the *Straits Times* stated that "whatever local industries it may develop in future, Singapore is today, and will always remain, primarily a transshipment port".[20] Along the same line, and until the early 1950s, the British Government did not seriously foster large-scale industrialization in the colony of Hong Kong. It was used mainly as an imperial outpost in the insular trade between China, India and Britain. Some complimentary industrial activities did develop, including shipbuilding, match manufacturing and repair shops, and provide some early experience in modern industries, it was trade that formed the mainstay of the economy.[21]

DIFFUSION OF WESTERN INDUSTRIAL TECHNOLOGY

While the British imperialists claimed that they had opened up the non-Western world to such blessings of Western technological civilization as railways, telegraphs and sanitation, the basic tenet observed by colonial administrators in the periphery was that technology transfers which might led to the growth of import-substitution industries were generally viewed with suspicion.[22] Colonial officials took the view that the development of such local industries would not only compete with manufacturers back home but "they threatened to bring forth native industrialists, engineers, technicians, and factory workers who would have challenged the authority of the colonial regimes".[23]

As with other dependent colonies in the British Empire, the colony of Singapore benefited from the introduction of modern technological systems, such as in transportation, communication and sanitation. Studies on the linkage between the city and diffusion of technological innovations have consistently pointed to the fact that the city played a leading role in the creation and diffusion of technological innovations which, in turn, hasten the process of urbanization and growth of the community.[24] Within the city, proximity and interpersonal contacts motivated the diffusion and acceptance of household-related innovations (such as the telephone and refrigerator) in family households which had the economic means to own such technological appliances.[25] By 1911, Singapore had developed into the most cosmopolitan city in Asia, with more than 185,000 — an increase of nearly 40 per cent from the 1870s.²⁶ For those locals, particularly the Straits Chinese, who had an enthusiasm for European fare and were able to afford them, the modern technological gadgets and appliances enhanced their living comforts and class status. Besides the Austin six-seater or Model "T" Ford, electric lighting and fans, the kerosene (later operated by electricity) refrigerator, filters, sewing machines, the wireless, the gramophone and silver tableware were added to the household to provide a sense of imperial grandiose. While household-related innovations found ready acceptance by the locals, it is not readily so in the case of business-related innovations, such as industrial machinery.

The transfer of industrial and production technology was carried out within the framework of British colonialism, that is, to increase the production of tropical products and their export to Britain in order to improve balance of payments and the tax base of the colony. There was no shortage of investment capital from British companies in the dependent colonies, including the Federation of Malay States and the Straits Settlements. British investment in Malaya from the 1870s until the 1960s was overwhelmingly concentrated in the tin, rubber and palm oil industry, accounting for about 93 per cent of the estimated total of 108 million pounds sterling in 1930.²⁷ As early as the 1880s, Singapore was, according to the British official N.B. Denny, "not only relatively but absolutely the richest of all the Crown Colonies under the British flag".²⁸ Also, Singapore (and Malaya) was not lacking in European and Chinese entrepreneurs who had the capital and the trading network to invest in Western technology and venture into manufacturing.²⁹ Why then British or, more generally, Western technology, failed to flow into the production and skilling of indigenous companies, particularly those involved in the rubber, mining and port businesses?

Importing British-made machinery was certainly the most cost-effective way of starting an industry but it required the entrepreneur to be technologically daring. And from the last decade of the nineteenth century, the world saw the introduction of a cluster of innovations that marked the start of a new upswing, popularly described as the Second Industrial Revolution.³⁰ However, to create what Alfred Chandler described as a "modern industrial enterprise", three sets of investments were needed, "investment in production large enough to utilize the economies of scale and scope inherent in technological innovation … in marketing and distribution large enough to sell the good produced by the new processes of production [and] in recruitment of a managerial hierarchy to manage and coordinate the day-to-day processes of production and distribution and to allocate resources for future production and distribution".[31] Even for British entrepreneurs who were "first movers' in tapping into the technological innovations in industries such as chemicals, machinery and mining, they failed to develop the competitive capacities and allowed the Americans and Germans to outperform them in many of these new industries.[32] Clearly, it was even more remote for the many migrants (largely Chinese) who were imbued with the desire to make their quick fortunes overseas, to become industrial entrepreneurs. Besides the heavy investment in buying modern technology and the access to international markets, one has to look into developing expertise in production, distribution and management. Unlike Hong Kong entrepreneurs, "there was meagre injection of immigrant capital and know-how into Singapore" and hence "a low degree of industrial capitalism among the leading Chinese businessmen".[33] On the other hand, the overseas Chinese networks within the Southeast Asian region favoured the intermediary or "middle-man" trading role because of the established *guanxi* or relationships amongst business associates and kin.[34] In any case, not all migrants were innovative and risk-taking and only a handful became successful entrepreneur.[35]

Perhaps the most prominent of these entrepreneurs was Tan Kah Kee. A shrewd, far-sighted and enterprising native of Fukien province in China, Tan made the best use of the favourable economic conditions in Singapore during the early decades of the twentieth century.[36] Recognizing the economic benefits of producing rubber goods and the use of his rubber manufacturing plants as training schools for Chinese skilled workers and technicians (who could ultimately played a role in industrializing China), Tan invested heavily in European technology by buying new and second-hand machinery and employing European production technicians.[37] He was also a businessman blessed with the creativity of an inventor; between 1924 and 1932 he took out numerous patent rights from the government.[38] The manufacturing of rubber shoes was a key industry, owned and managed by Chinese businessmen. During the early 1930s, and together with Hong Kong manufacturers, they vigorously expanded their export of cheap rubber shoes to Canada and Britain, resulting in strong protest by the Canadian and British manufacturers. In 1935, the largest rubber shoes factory owned by Tan Kah Kee went bankrupt and this allowed Hong Kong firms to penetrate further the British market.[39] Interestingly, the trading mentality of the Chinese community was likely to be enhanced by similarities in terms of business operations between British and Chinese firms in Singapore. As in Britain, small businesses dominated the Singapore economy. There was a lack of vertical integration in business. Separation of ownership from management was not commonly practised, and the concept of limited liability was not popular. In both countries, most businesses remained family-linked partnerships though mergers between British firms did take place in the early decades of the present century.[40] Hence, inheritance and continuation of trading businesses was a common practice in colonial Singapore, as in the case of prominent business leaders like Teo Lee, Tan Kim Seng, Hoo Ah Kay, Wee Ah Hood, Cheang Hong Lim who were second-generation wealthy Chinese.[41]

For the colonial government, the official perspective for the lack of concerted effort in transferring industrial technology could be simplified down to the fact that Singapore was viewed nothing more than an entrepôt and naval station and not as an extension of Britain's industrial production machinery in the Far East. For the local businessmen, institutional and socio-cultural factors inhibited the rise of a technological or manufacturing culture during the period of British colonial rule.[42] Relations between British firms and Chinese firms were dominated entirely by commerce. In Inkster's words, "much of the problem of technology transfer into Singapore is explainable in terms of entrepôt colonialism" in which "only selected technique entered the system in the first place, and that British activity was focused on the three initial building blocks of banking, British engineering shops and firms concentrated on repair and services, and successful port building and shipping".[43] While the British managing agencies and British engineering companies had direct access to the industrial emporiums in Britain, imported machinery used in the mining industry was sold largely to British mining and engineering companies — and not to local Chinese business community. There was simply insufficient or total lack of demand for such machinery. In any case, besides the tin-smelting facilities at Pulau Brani, there were hardly any tin-mines or extensive rubber plantations in Singapore. Some historians have argued that British managing firms were:

> notoriously poor at effectively transferring the advanced knowledge, skills and technology at their disposal to the inhabitants of the colonies where they invested. They did not provide training facilities or employ indigenes at any but the most unskilled tasks ... The colonial governments sometimes criticised expatriate firms for this practice, but in reality were

themselves no more willing to train and employ indigenes in responsible positions ... Indeed, for most of the period they decried the emergence of the "educated native". Thus colonialism denied these countries one of the major potential benefits of direct foreign investment, the transfer of technology and skills and the stimulation which the activity of the foreign firm in the host economy can give to local capitalists.[44]

However, in the case of colonial Singapore, the transfer of British maritime technology and technical skills to local shipping companies was significant and successful. As mentioned earlier, Bun Hin (Green Funnel), Wee Bin & Co. and the Straits Steamship Company purchased mainly second-hand British-made ships as the bulwark of their coastal fleet.

One can argue that the commercial success of the city-port also served as a hindrance to the acquisition of foreign industrial technologies as an essential stage towards the development of an extensive indigenous manufacturing base. The thousands of immigrants who came to Singapore had one ulterior motive — to carve out a niche and make a fortune. Unlike the British, however, the Chinese who (came largely from South China) owed their loyalty "to trading opportunity, to those who traded fairly with them, and to stable regimes which protected their concerns".[45] The colonial governors gave to the Chinese precisely what they were looking for. As in Hong Kong, where the link between the "political supremacy of the British and the economic indispensability of the Chinese" was a dynamic component of the trading city-state, the British in Singapore also permitted the Chinese to become economically independent and indispensable.[46] Why was it the case? Simply put, the Chinese had been in Southeast Asia long enough to know the "tricks of the trade". From the days of Francis Light and Stamford Raffles, and certainly way before, the diaspora of Chinese immigrants to Southeast Asia was closely associated with their trading and agricultural activities throughout the region. They had adapted to the environment and acquired a working knowledge of the local languages used. As Raffles foresaw, due to its strategic position, the greatness of Singapore would be largely dependent on its role as the middleman of the region's trade. Naturally, to the Europeans, the Chinese were seen as the best candidates to act as middlemen. To the many British folks who came to this part of the world, and especially for the first-timers who came to this tropical frontier to seek economic prospects and fortune, the Chinese provided useful bridging services between themselves and an alien environment of different cultures and languages. In the nineteenth century, government revenues also depended heavily on the economic activities of the Chinese, especially for their investments in the tin and rubber industry, and British officials relied on Chinese *kapitans* and other Asian leaders in controlling the local population.[47] Until the early 1930s when immigration quotas were introduced, the colonial government also recognized the importance of the mass influx of unskilled Chinese workers to the development of the port and the basic infrastructure of the town. As Singapore prospered, the collaboration between the British and Chinese entrepreneurs, especially the English-educated Straits-born Chinese, became more pronounced. To further strengthen the collaboration between the ruler and the ruled, and as an indication of respect, prominent Chinese leaders like Lim Boon Keng, Tan Kah Kee, and Tan Lark Sye were co-opted to become unofficial members in the Legislative Council of the Straits Settlements. There existed a kind of division of labour — the British officials administered the colonial affairs, most of the Europeans were either managers or technical experts, and the Chinese involved themselves in their own comprador activities, small businesses and menial jobs.

On the whole, the Chinese were left to run their own affairs and although the community itself was fragmented by the many dialects and clan organizations, a sense of unity was achieved through the operations of their various types of associations.[48] They also brought along with them the Confucian hierarchical order expressed in class, clan and family structures. Outside traditional China, wealthy merchants ranked at the top, followed, in descending order, by a handful of educated elites, the artisans and the large body of coolies.[49] An exhibition of wealth would lead to a showering of respect and was regarded as an indication of high social status. Therefore, the deep desire to gain a respectable social status was seen as an incentive for overseas Chinese to acquire wealth — and to acquire it as quickly as possible. Aligning with Singapore's economic role, trade and the operation of small businesses, not the production of manufactured goods was seen as the quickest, safest and most viable path to the accumulation of wealth.

INSTITUTIONAL KNOWLEDGE: GROWTH OF CHINESE BANKING

While the colonial administration and the European agency houses did not actively pursue a policy of importing industrial technology to develop a manufacturing sector, technology transfer in the form of institutional knowledge and skills did contribute to the strengthening of the Singapore's economy. Local Chinese merchants and compradors formed an important structural link in Singapore's entrepôt trading framework. Operating under a *laissez-faire* policy and enjoying the law and order of *Pax Britannica*, Chinese merchants learnt "to adapt to new methods of operation, to become more innovative themselves and to use the favourable parts of foreign cultures to help them modernize their minds as well as their institutions".[50] The dynamics of the interlocking trading network between British firms and Chinese compradors allowed many learning situations where new knowledge and skills were transferred to the locals.

Following a period of acclimatization by the Chinese to the ways the British operated its commercial system, one of the most significant results of these learning situations was the rise of Chinese banking in Singapore. The three British banks which opened their doors in the second half of the nineteenth century — the Chartered Bank, the Mercantile Bank, and the Hong Kong and Shanghai Banking Corporation — maintained British banking tradition of concerning mainly with advances, overdrafts, and servicing international trade with exchange bills.[51] Western banks played an indispensable role in helping the smaller traders with the essential start-up capital. They recognized the useful services provided by Chinese compradors, book-keepers and clerks within their organizations and loaned money to Tamil *chettiars*, who in turn provided credit to small-scale Asian traders, artisans, tin-miners and smallholders.[52] However, according to Landes, from a historical perspective, British banks' greatest weakness "which became apparent only after the middle of the nineteenth century, was its inability to initiate or encourage the kind of industrial enterprise that would call for large amounts of outside capital".[53]

The British banking model was largely adopted by the Chinese in their own ventures to establish Chinese-owned banks. Hitherto the Chinese in Singapore were not accustomed to develop modern capitalist organization or banking. Many continued to operate within the traditional family linkages and personal form of trading. However, a few open-minded entrepreneurs were more than willing to adapt themselves to the new Western capitalism. These newly formed Chinese banks also gained the invaluable services and skills of former Chinese employees of British banks. Although constantly supported by the goodwill of British banks, Chinese banks in Singapore did not quite enjoy the growth and stability of their European counterparts mainly due to inexperience, malpractice, excessive conservatism and clan rivalry.[54] Nevertheless, the learning and acceptance of British commercial and banking practices did take place, and gradually the Chinese population became convinced of the desirability of keeping their savings in banks. In turn, the Chinese banks became important mechanism for the modernization of Chinese business practices through the availability of loans to Chinese entrepreneurs and shopkeepers.[55] But the transfer and diffusion of British banking practices could actually work against the rise of large-scale industrialization in the colony. The Chinese banks adopted the practice of "short-termism" of British banks, that is, of providing only short-term loans. The lack of long-term financial support discouraged merchants from diversifying into manufacturing interests.[56] Chinese capital remained largely engaged in comprador-type of business ventures which they were most familiar. It is also possible that the close network of commercial dealings within a small geographical setting led to an assimilation of some British business cultural traits to Chinese firms. According to the Stanworths' study, traditional British managers had a narrow view of corporate culture. They were sceptical about technical education and new ideas in management specialisms, such as marketing and manpower planning. Instead of relying on specialists, British firms frequently nurtured their managers through the rank-and-file.[57] There was also the paternalistic management approach in order to develop a compliant and co-operative labour force.[58] Another cultural legacy of British business was an emphasis on "fairness" — what the British commonly referred to as the "culture of the silent majority". Presumably, it promoted a sense of fair play, of playing the game rather than winning, and of leaving behind something to the new generations. There was also the culture of complacency about competition in British firms, mainly due to a historical reliance on imperial markets.[59] Admittedly, it is difficult to state conclusively the extent to which these British work ethics were diffused into Chinese business organizations. But what seemed rational to the Chinese, as to the British, was the use of indigenous capital on short-term, speculative trading opportunities, preventing its widespread utilization in the industrial sector. Perhaps this was what Wang Gungwu refers to as the "risk-taking features" of Chinese entrepreneurship — not just boldness in risk-taking but a combination of qualities of "industriousness, shrewdness, ruthlessness, something that might be called the will to profit".[60] In short, for the majority of the Chinese merchants in colonial Singapore, manufacturing ventures that required the importation of foreign technology, heavy capital outlay and long-term planning were not eagerly pursued. Those who possessed some skills and a small capital usually became traders, subcontractors or shopkeepers. Clearly, it was not easy to shake off this trading and brokerage mentality and to invest in modern technology and manufacturing activities. Right up to the 1960s, manufacturing share of the labour force ranged from 10 per cent to 15 per cent, while its contribution to domestic income stagnated at 5 per cent to 7 per cent.[61] The entrepôt trade remained as the main engine of growth.

Thus, within Singapore's trading framework, an unspoken and interdependent tripartite partnership between the government, the European merchants and the Chinese existed since the early decades of the nineteenth century. This partnership produced wealth and motivated the indigenous population to continue their various roles within the trading network. The industrial enterprises started by the Chinese occurred largely without much support from or impediments imposed by the colonial administration. British business flourished in Singapore and Malaya because the colonial administration generally adopted a pro-business policy. And although the British remained socially exclusive and segregated from the masses, the business partnership withstood the stress and test of time. British officials and European traders alike did not wish to jeopardize the social relationship. Indeed, most Europeans consciously promoted racial harmony in order to uphold their reputation as the civilised, ruling race.[62] As one contemporary observer noted, "A white man in the tropics is watched as a minor god.... From the minute he has entered his office, he is watched by natives and feels, if he has a conscience, that he must give an example, in conduct, wisdom and strength".[63]

Notes

1. Headrick, *Tentacles of Progress*; M. Havinden and D. Meredith, *Colonialism and Development: Britain and its Tropical Colonies, 1850–1960* (London: Routledge, 1993); and David Meredith, "The British Government and Colonial Economic Policy 1919–1939", *Economic History Review* 28 (1975): 484–99.
2. Tomp Kemp, *Industrialization in the Non-Western World* (London: Longman, 1983), p. 5; Headrick, *Tentacles of Progress*, p. 16.
3. Huff, *Economic Growth of Singapore*, p. 208.
4. John H. Drabble, "Technology Transfer in Singapore/Malaya during the Colonial Period", *Journal of the Malaysian Branch of the Royal Asiatic Society* LXXVI (2003): 82.

5. Headrick, *Tentacles of Progress*, p. 4.
6. Colony of Singapore, *Report of Commission on Trade of the Colony 1933–34* (Singapore: Government Printing Press, 1934), p. 146.
7. Huff, *Economic Growth of Singapore*, p. 225.
8. *Report of Commission on Trade of the Colony*, p. 146.
9. K.G. Tregonning, *Straits Tin: A Brief Account of the first Twenty-Five Years of the Straits Trading Company, Limited, 1887–1962* (Singapore: The Straits Times Press, n.d.), pp. 25 and 44.
10. See Yip Yat Hoong, *The Development of the Tin Mining Industry of Malaya* (Kuala Lumpur: University of Malaya Press, 1969). For a recent comprehensive study of the role of the tin industry in Malaysia's economic development, see Salma Nasution Khoo, *Kinta Valley: Pioneering Malaysia's Modern Development* (Perak, Malaysia: Perak Academy, 2005).
11. A. Wright and H.A. Cartwright, eds., *Twentieth Century Impressions of British Malaya: Its History, People, Commerce, Industries, and Resources* (London, 1908; reprinted Singapore: Graham Brash, 1989), p. 212. During the early years of tin mining in the Malay States, no machinery of any kind was used beyond the Chinese endless wooden chain pump and the overshot water wheel — modelled after irrigation techniques used by rice paddy farmers in China. In the era of the "Gold Rush" in California during the 1850s, the Chinese introduced the water wheel to American placer mining. This device allowed the miners to pump and sluice water from the river, which was then used to wash gravel from gold. The pumping method was not only derived from Chinese agriculture, but from generations of experience from tin miners in Guangdong, who had in turn acquired their knowledge from Chinese miners in Malaya.
12. "Malaya's Merchant Houses", *British Malaya*, November 1937, pp. 173–74.
13. Drabble, *"Technology Transfer"*, pp. 82–83; Huff, *Economic Growth of Singapore*, pp. 22, 61, and 214.
14. See Norman Miners, "Industrial Development in the Colonial Empire and the Imperial Economic Conference a Ottawa 1932", *Journal of Imperial and Commonwealth History* 30, no. 2 (2002). Miners argues the generalization that the development of manufacturing industry in the colonies was very limited does not apply to Hong Kong. Hong Kong entrepreneurs had been building up the manufacturing sector since the nineteenth century, and by 1939 the colony had "built up a flourishing export trade in manufactured goods to China and neighbouring Asian countries and was even successfully competing with British firms in a few items in the British home market" (p. 54). See also Ngo Tak-Wing, "Industrial History and the Artifice of Laissez-Faire Colonialism", in *Hong Kong's History: State and Society Under Colonial Rule*, edited by Ngo Tak-Wing (London and New York: Routledge,

1999), Chapter 7. Ngo too argues that "the supposedly 'transforming role' of colonial rule is largely the product of a one-sided account of Hong Kong's development trajectory given by the dominant historiography.... All signs point to the fact that the manufacturing industry had become a significant economic activity by the 1930s" (pp. 119 and 126).
15. *Report of Commission on Trade of the Colony*, pp. 147–50.
16. Ibid., pp. 147–48.
17. Ibid., p. 148.
18. Ibid., pp. 151–52.
19. Ibid., p. 151.
20. Quoted in *British Malaya*, March 1929, p. 305.
21. Steve Tsang, *A Modern History of Hong Kong* (London and New York: I.B. Tauris, 2004), p. 62.
22. Headrick, *Tentacles of Progress*, p. 381.
23. Ibid.
24. Paul Bairoch, "The City and Technological Innovation", in *Favorites of Fortune: Technology, Growth, and Economic Development since the Industrial Revolution*, edited by Higonnet, P., David S. Landes and Henry Rosovsky (Cambridge, Massachusetts: Harvard University Press, 1991), pp. 159 and 172. Bairoch provided a review of some key studies done on the role of cities in stimulating and diffusion innovations.
25. Ibid., pp. 164–72.
26. Turnbull, *History of Singapore*, p. 95.
27. Junid Saham, *British Industrial Investment in Malaya 1963–1971* (Kuala Lumpur, 1980), pp. 18-19.
28. Comments made by N.B. Denny in talk given by Sir Frederick Weld, quoted in Kratoska, *Honourable Intentions*, p. 82.
29. In the case of Malaya, see Wong Yee Tuan, "More Than a Tea Planter: Jon Archibald Russell and His Businesses in Malaya, 1899–1933", *Journal of the Malaysian Branch of the Royal Asiatic Society* 83, Part 1 (2010): 29–51. The popular perception is that British agency houses were largely instrumental in charting the economic development of Malaya during the early decades of the twentieth century. Wong's biographical study aims to illustrate his argument is that local British entrepreneurs too played critical roles.
30. D.S. Landes, *The Unbound Prometheus: Technological Change and Industrial Development in Western Europe from 1750 to Present* (Cambridge: Cambridge University Press, 1969), p. 235. In pp. 326–58, Landes offers some answers to the question: Why did industrial leadership pass in the closing decades of the nineteenth century from Britain to Germany. The answer lies strongly in cultural factors.
31. Alfred D. Chandler Jr., "Creating Competitive Capability: Innovation and

Investment in the United States, Great Britain, and Germany from the 1870s to World War I", in *Favorites of Fortune*, edited by Higonnet, Landes and Rosovsky, p. 432.
32. Ibid., pp. 454–56.
33. Chan K.B. and Claire Chiang S.N., *Stepping Out: The Making of Chinese Entrepreneurs* (Singapore: Prentice Hall, 1994), p. 33.
34. Ibid.
35. Ibid., p. 35.
36. The authoritative work on Tan Kah Kee is Yong Ching Fatt, *Tah Kah Kee: The Making of an Overseas Chinese Legend* (Singapore: Oxford University Press, 1987). Using mainly Chinese primary sources found in Singapore and Fukien, Yong produced a meticulous study of the famous Chinese entrepreneur, philanthropist and educationist.
37. Ibid., p. 56.
38. Ibid., pp. 58–59.
39. At its height in 1929, Tan Kah Kee's enterprise employed 4,000 workers who churned out 20,000 pairs of shoes a day. The collapse of rubber prices and the onset of world depression affected severely his manufacturing venture.
40. Mansel G. Blackford, *The Rise of Modern Business in Great Britain, the United States, and Japan* (Chapel Hill: University of North Carolina Press, 1988), pp. 65–68.
41. Song Ong Siang, *One Hundred Years' History of the Chinese in Singapore* (London: John Murray, reprinted, Singapore: Oxford University Press, 1991), pp. 33, 46, 52, 102 and 168.
42. See Goh Chor Boon, "Imported Technology", *Journal of the Malaysian Branch of the Royal Asiatic Society* LXXI (1998): 41–54 and the response to this article in I. Inkster, "The Trouble with Technology: Comments on the Experience of Singapore under Entrepôt Colonialism", *Journal of the Malaysian Branch of the Royal Asiatic Society* LXXIII (2000): 107–15. Inkster develops a useful model to denote the potential linkages between the various sectors of the Singapore (and Malaya) colonial economy and to explain why there a general lack of technology transfer from the European agents (mainly trading houses, banks, engineers and colonial botanists) into the indigenous agricultural and manufacturing (more specifically, the mining) industry. Both authors framed their arguments in terms of domestic institutional and socio-cultural constraints brought about by the development of Singapore as an entrepôt. I would like to thank Professor Inkster for his invaluable comments.
43. Inkster, "The Trouble with Technology", pp. 109 and 114.
44. Havinden and Meredith, *Colonialism and Development*, p. 304.

45. Wang Gungwu, *China and the Chinese Overseas* (Singapore: Times Academic Press, 1991), p. 170.
46. Chan Wai Kwan, *The Making of Hong Kong Society: Three Case Studies of Class Formation in Early Hong Kong* (Oxford: Clarendon Press, 1991), pp. 10–11. Chan's study examines the sociological origins of the labour movement and the emergence of two complementary elite groups, the British merchants and the Chinese traders in pre-war Hong Kong. Using case studies, Chan attributes the existence of the two elite groups to two main aspects, first, the racial and cultural differences and, second, the political power of the British and the economic strength of the Chinese.
47. J.G. Butcher, *The British in Malaya, 1880–1941: The Social History of a European Community in Colonial Southeast Asia* (Kuala Lumpur: Oxford University Press, 1979), p. 226.
48. Yen Ching Hwang, *A Social History of the Chinese in Singapore and Malaya, 1800–1911* (Singapore: Oxford University Press, 1986), p. 318; Yong, *Tan Kah Kee*, p. 12.
49. Yen, *Social History of the Chinese*, p. 319.
50. Wang, *Chinese Overseas*, p. 195.
51. Lim Chong Yah, *Economic Development of Modern Malaya* (Kuala Lumpur: Oxford University Press, 1967, reprinted 1969), p. 116.
52. The term "chettiar" is an Anglicized form denoting Tamils in Southeast Asia operating as money-lenders.
53. Landes, *The Unbound Prometheus*, p. 349.
54. Tan Ee Leong, "The Chinese Banks Incorporated in Singapore and Malaya", in *Readings in Malayan Economics*, edited by Thomas H. Silcock (Singapore: Eastern Universities Press, 1961), p. 456; Yap Pheng Gek, *Scholar, Banker, Gentleman Soldier: The Reminiscences of Dr Yap Pheng Gek* (Singapore: Times Books, 1982), p. 35.
55. Edwin Lee, "The Colonial Legacy", in *Management of Success*, edited by Sandhu and Wheatley, Chapter 1.
56. John Stanworth and Cecilia Stanworth, *Work 2000: The Future for Industry, Employment and Society* (London: Paul Chapman, 1991), p. 231. According to this study involving in-depth interviews with twenty-six prominent individuals from different background, the "short-termism" policy of British banks is still prevailing today, much to the competitive disadvantage of Britain. The researchers found that many of British industries and firms are still clinging on to their traditional, archaic work ethics which, according to the authors, actually served to erode their competitiveness.
57. Ibid., p. 179.
58. Ibid., p. 182.
59. Ibid., p. 183.

60. Wang, *The Chinese Overseas*, p. 195.
61. Iain Buchanan, *Singapore in Southeast Asia: An Economic and Political Appraisal* (London: G. Bell and Sons, 1972), p. 35.
62. Butcher, *British in Malaya*, pp. 167–68.
63. George Bilainkin, *Hail Penang! Being the Narrative of Comedies and Tragedies in a Tropical Outpost, among Europeans, Chinese, Malays, and Indians* (London: Sampson Low, Marston and Company, 1932), p. 69.

Conclusion

Without the aid of technological implements and scientific knowledge, vast areas of the world not have been subjugated by Britain. But technology was not a cause of imperialism. It was a facilitator of imperialism. British engineers proceeded to alter the landscape of the imperialized world with bridges, railways, roads, tunnels, irrigation and sanitation systems, harbours, telegraph lines, submarine cables and electrification of the urban environment. The transformation of Singapore from a fishing village to a modern, bustling commercial city, technologically linked to world markets and protected by an elaborately constructed naval base in the Far East is undoubtedly an impressive achievement of Britain's "civilising mission". As the port and city grew, modern scientific and technological systems were introduced for the benefits of the Europeans in the first instance and later diffused to the masses at large. Modern communications and transport technology led to large declines in costs of travelling and transportation of goods. The telegraphy and telephone linked the small domestic market in Singapore to the rest of the world and the availability of electrical power after 1940 must be seen as a large net benefit to society and economy after 1940. In the words of David Cannadine, "these technological transformations were intrinsically significant as the agents and avatars of imperial modernity rather than imperial conservatism; they were, in addition, the harbingers of social developments and political changes..."[1] While the Singapore colonial government failed to provide sufficient housing for the masses, it did work hard to improve the overall living standards and material welfare of the people. For the local folks who could afford British products of modern living made possible through the application of science and technology — refrigerators, Holloway's Pills and Ointments, Pears soap, Cow & Gate milk powder, face towels, Singer sewing machines, machine-made sewing needles, perfumery and cosmetics, knitted stockings, glasswares, oil lamps, mechanical clocks, motor car, etc — life was good and comfortable. These "imperial" commodities, which were consumed even at the frontiers of the known world, were a measure of the global reach of the values of Britain's commercial civilization.

Contemporary writings of former British officials who had served in Singapore frequently pointed out the dramatic changes, including the entrenchment of British institutions and ideas that accompanied the transformation of the trading town to a bustling metropolis. The British in Singapore ruled a trading port through the imposition of British institutions relating to private property, the rule of law, education in Western knowledge, health and social welfare and the liberty of the individual. An intangible and lasting legacy of British colonialism is the ideal of a good, uncorrupted government, supported by an equally efficient bureaucratic system. Although top-level administration was in the hands of a small group of Englishmen, the functions of the Straits Civil Service, created in 1934, were instrumental in ensuring civil peace and order. The overriding political philosophy of the government and the civil service was to "hold aloft the banner of justice, truth, and right-dealing".[2] John Davis, who served in the police force in the 1930s, commented: "The whole basis of our education was that we should always remain completely incorruptible. One of the great points about the British Empire was that although we were powerful, difficult, bossy people who may have been immensely disliked by the inhabitants of the countries which we ruled, they accepted it simply because they knew we were not involved. Very, very few British people would be involved in corruption because, even if we had an inclination towards being corrupt, there was nothing really that they could offer us that would have had any great significance".[3] Undeniably, this legacy became a sacrosanct practice for all civil servants in Singapore today and is one of the great pillars of its public life.

But the British were never systematic Anglicizers. For all their rhetoric about civilizing the "backward" peoples of the Orient, colonial administrators were generally "reluctant spenders, ever fearful that tampering with native laws, faiths, or learned traditions might undermine their fragile authority over large and often turbulent subject populations".[4] British officials who planned and administered colonial cities like Singapore had to do so within the overall framework of colonialism. The paternalistic approach stressed on the maintenance of law and order, rather than the progressive uplift of the colonized. The basic tenet was, of course, the preservation of British power and prestige. Technology served ideological as well as physical purposes. It was directed at the colonizers as at objective prove of superiority and a justification for white domination. Thus, social engineering and town planning were carried out simultaneously such that rapid urbanization and a healthy environment would be created for the benefits of the Europeans.[5]

Unlike the case of the American colonization of the Philippines where "[f]rom the outset, it was assumed that engineers would take a leading role [and] set the standards for Americans engaged in the civilizing mission", British and European businessmen and entrepreneurs were the more prominent drivers of change than their engineering counterparts in a rapidly modernizing Singapore city before the Second World War.[6] These change agents had little difficulty in establishing and expanding their business operations. There was no war of subjugation of a hostile, indigenous population. There was no displacement of cottage industries by a more advanced technological culture. There was no serious clash of interest and no cause of tension between the merchant imperialists and the Asian merchants. There were hardly any natural resources for the British imperial merchants and administrators to exploit. Even rubber and tin, boosted by the transfer of science and modern machinery by British firms, were largely processed and mined in colonial Malaya. If anything, what was "exploited" was the willingness and work ethics of the Chinese — as middlemen par excellence and general workers — to develop a collaborative, trading network with their Western counterparts. Just like the poor English, Scottish and Irish indentured migrants who first arrived in the New World at Chesapeake Bay off the coast of what today is South Carolina in 1670s, the Chinese came in droves from southern China to find a better life in Singapore. There was even collaboration between the local elites (mainly the Chinese) and the British colonial statesmen, best exemplified by the Queen's Scholars, Lim Boon Keng and Song Ong Siang who finished their tertiary education in England in the 1890s and were exposed to the philosophy of the Utilitarians and Evangelicals. But these Chinese elites were neither mindless imperialist tools nor blind to their own interests. In all sincerity, they collaborated because they recognized the fairness, strength and power of British principles of justice, and British ideas of progress.⁷ Indeed, more often than not, they could influence colonial policies by wielding their own literary expertise, capital resources, local command structures and networks of production and distribution. As for the British emigrants — from the businessmen to the port engineers — they had, according to Peter Marshall, "a deliberate, sustained and self-conscious attempt to order, fashion and comprehend their imperial society overseas on the basis of what they believed to be the ordering of their metropolitan society at home".⁸ More often than not, they had a good life out in tropical Singapore — living in comfortable suburban homes, enjoying the wide range of imported consumer goods, and saving enough for a well-earned retirement back in homeland. Although industrial technology transfer was limited and its diffusion largely blocked by social and institutional factors inherent in economic and cultural milieu of the Singapore's colonial society, these European agents of change opened up opportunities, however limited, for some cultural diffusion of modern technical and management knowledge and skills to indigenous entrepreneurs and skilled workers. There was also no Luddite-like response from the masses to the gradual adjustments of their lifestyles as a result of modern technology. If not for the lack of the provision of English and technical education by the colonial government, more would have learned about the benefits of Western science and technology. In short, British colonialism, imported technologies and the influx of Chinese immigrants into the island (and Southeast Asia as a region) created the right concoction that would eventually produce a cosmopolitan society. By the end of the nineteenth century, the colony had matured into a commercial emporium — a staple entrepôt for the Malaya hinterland — lubricated by its free-trade policy and endowed by its strategic location.

Finally, while it is true that the passing of time could lessen the impact of historical antecedents, a study of the complexity of technology transfer, colonialism and development would provide a useful hindsight and perspective into issues facing former empire colonies quest for scientific and technological excellence as an integral component of their national growth strategy. Although the issue is still a contentious one, South Korea and Taiwan were able to enjoy success in their technological leapfrogging strategies starting in the 1960s because Japanese colonialism had modernized the economy and society using Western technology, with construction of extensive networks of railroads, opening of ports and heavy investments in steel, chemicals, hydro-electric power and communication industries, and coupled with the expansion of literacy and technical training. Long exposure to industrial and engineering technology provided the Koreans and the Taiwanese invaluable learning experiences in adapting foreign technologies and, in the long run, in creating an indigenous, self-reliant technological base. In the case of Singapore, one of the most critical problems facing Singapore's current heavy investment (and mainly public-funded) in its research and development programmes is the acknowledged lack of indigenous technological entrepreneurs, research engineers and scientists who are willing to venture into high-tech, high value-added industries and research and development activities. Assimilation of imported or foreign technology is by no means easy mainly because, unlike normal goods, technology is not something just to be bought and consumed. An individual, a firm, or a country that had shaken off the yoke of imperialism and colonization, has to possess accumulated capabilities to understand it and apply it in its own environment which often differs significantly from the environment in which the technology was invented.[9] The extent and content of capabilities one has at a particular point in time depends on the path in which one has accumulated capabilities. This path-dependency — the tendency of a past or traditional practice to continue even if better alternatives are accessible — of a country's growth history inevitably influences its growth pattern. Entrepôt colonialism in Singapore had successfully transferred and nurtured commercial, financial and brokerage institutions and practices to local entrepreneurs. This comprador or brokerage service culture is ingrained within the business community which has produced many of Singapore's rich and successful individuals. Science and technology, on the other hand, are kept at a distance during the years of British rule. As late as the 1980s, government efforts to promote research and development activities were dampened by the lack of a critical mass of scientists and research engineers and limited research fund. It is impossible to understand Singapore's experience in creating a viable scientific community and a culture of cutting-edge research and development without reaching back into pages of Singapore's colonial history.

In 1945, when the British came back to reclaim Singapore, Britain was a bankrupt nation. There was no way for her to regain the technological supremacy she once so gloriously demonstrated as the "workshop of the world". While historians would mull over the loss of Singapore to the ineptness of the British Government, the chauvinism and the racial supremacy exhibited by the British, this book has illustrated how the entrepôt economy and the urbanization of the city was shaped by imported Western technologies and had benefited from the imposition of a set of norms, behaviours and institutions relating to the rule of law, environmental science, material living, business knowledge and work ethic. From the British administrators, engineers to the British businessmen, these change agents promoted a kind of imperial welfarism, a programme of development that, however ethnocentric and patronizing, was nonetheless well intended. Today, a high proportion of Westerners who live in Asia are located in Singapore or Hong Kong. Journalist Martin Jacques maintains that Hong Kong still "bears the colonial imprint, while Singapore, more than any other place in the region, has sought to make itself into the Asian home of Western multinationals, a kind of Little West in the heart of Asia.... The great majority [of the expatriates] live in a handful of salubrious, Western-style residential 'colonies', enjoying a life of some privilege, such that for the most part they are thoroughly insulated from the host community...".[10] Indeed, "colonial style" makes a reappearance in interior design and fashion, alongside tropical collections. It is as if people (not just the European expatriates) are reliving the fantasies of what the late Mary Turnbull described as the "High Noon of Empire", out at the imperial frontier.

Notes

1. David Cannadine, *Making History Now and Then: Discoveries, Controversies and Explorations* (New York: Palgrave Macmillan, 2008), p. 149.
2. Comments by William Adamson in talk given by Sir Frederick Weld, quoted in Paul Kratoska, *Honourable Intentions: Talks on the British Empire in South-East Asia Delivered at the Royal Colonial Institute, 1874–1928* (Singapore: Oxford University Press, 1983), p. 77.
3. Quoted in Allen, *Plain Tales from the British Empire*, p. 626.
4. Susan Bayly, "The Evolution of Colonial Cultures: Nineteenth-Century Asia", in *The Oxford History of the British Empire; The Nineteenth Century*, edited by Andrew Porter (New York and Oxford: Oxford University Press, 1999), Chapter 20, p. 450.
5. P.J. Christopher, *The British Empire at Its Zenith* (London: Croom Helm, 1988), p. 131.
6. Quotes by Adas, *Dominance by Design*, pp. 144 and 146.
7. Lee, *The British as Rulers*, p. 290.

8. P.J. Marshall, "Empire and Authority in the Later Eighteenth Century", *Journal of Imperial and Commonwealth History* 15 (1987): 105.
9. For accounts and explanations of how the Japanese, the Koreans and the Taiwanese became successful technological innovators, see Hiroyuki Odagiri and Akira Goto, *Technology and Industrial Development in Japan: Building Capabilities by Learning, Innovation and Public Policy* (Oxford: Clarendon Press, 1996) and Tessa Morris-Suzuki, *The Technological Transformation of Japan: From the Seventeenth to the Twenty-first Century* (Cambridge: Cambridge University Press, 1994), Dan Breznitz, *Innovation and the State: Political Choice and Strategies for Growth in Israel, Taiwan and Ireland* (New Haven and London: Yale University Press, 2007), Ezra Vogel, *The Four Little Dragons: The Spread of Industrialization in East Asia* (Cambridge: Harvard University Press, 1991).
10. Martin Jacques, *When China Rules the World* (London: Allen Lane, 2009), p. 101.

Bibliography

Unpublished Official Records
Administrative Reports of the Singapore Municipality, 1899 and 1900.
Straits Settlements Annual Reports, 1855–1912.
Straits Settlements Legislative Council Proceedings, 1867–1941.

Published Official Records
Report of the Commission of inquiry into the System of English Education in the Colony. Singapore: Government Printing Office, 1902.
Report on the experiment tapping of para rubber trees in the Botanic Gardens, Singapore, for the Year 1904. Singapore: Government Printing Office, 1905.
Report of technical committee. Singapore: Government Printing Office, 1926.
British Empire Exhibition Pamphlets, Series Nos. 1–9, 1928.
Report of the Telegraph and Telephone Communications Committee, 31 August 1931. Singapore: Government Printing Office, 1931.
Report of Commission on the Trade of the Colony, 1933–34, Vol. 1. Singapore: Government Printing Office, 1934.
Report of the Committee appointed by His Excellency the Governor of the Straits Settlements to Enquire into the Report on the Present Traffic Conditions in the Town of Singapore, 29 August 1938. Singapore: Government Printing Office, 1938.
Report on education in engineering by Professor G. McOwan. Singapore: Government Printing Office, 1938.

Personal Papers
O'Grady, G. J. *Being a True and Unbiased Account of the Life and Work of a Civil Engineer in Malaya, February 1928–February 1942*, Mss. Ind. Ocn, r6, 1942 (Typescript).

Unpublished Thesis

Ng Yeen Chern. "Nutrition and Health: Milk in British Malaya". Unpublished honours thesis. Department of History, National University of Singapore, 1999/2000.

Singh, Kuldip. "Municipal Sanitation in Singapore, 1887–1940". Unpublished honours thesis. Department of History, National University of Singapore, 1989.

Tan Chor Suan Donald. "Imperial Thoroughfares: A History of Roads in British Malaya". Unpublished honours thesis. Department of History, National University of Singapore, 2003/2004.

Newspapers

Singapore Free Press
Daily Advertiser
Straits Times
The Times

Books and Articles

Adas, Michael. *Machines as the Measures of Men: Science, Technology and Ideologies of Western Dominance*. Ithaca: Cornell University Press, 1989.

———. *Dominance by Design: Technological Imperatives and America's Civilizing Mission*. Cambridge and Massachusetts: The Belknap Press, 2006.

Aldcroft, D.H. "The Economy, Management and Foreign Competition". In *Where Did We Go Wrong? Industry Education and Economy of Victorian Britain*, edited by G. Roderick and M. Stephens, pp. 13–31. London: The Falmer Press, 1981.

Allen, Charles. *Plain Tales from the British Empire*. London: Abacus, 2008.

Allen, G.C. and A.G. Donnithorne. *Western Enterprises in Indonesia and Malaysian: A Study in Economic Development*. New York and London: Allen & Unwin, 1957.

Ambirajan, S. "Steam Intellect and the Raj: South India in the Nineteenth Century". In *The Steam Intellect Society*, edited by Ian Inskter, pp. 160–80. University of Nottingham: Department of Adult Education, 1985.

Armytage, W.H.G. *A Social History of Engineering*. Faber and Faber, 1976.

Baber, Zaheer. *The Science of Empire: Scientific Knowledge, Civilization, and Colonial Rule in India*. New York: State University of New York Press, 1996.

Bairoch, Paul. "The City and Technological Innovation". In *Favorites of Fortune: Technology, Growth, and Economic Development since the Industrial Revolution*, edited by P. Higonnet, David S. Landes and H. Rosovsky, Chapter 5. Cambridge, Massachusetts: Harvard University Press, 1991.

Baker, Nicholas. *Human Smoke: The Beginnings of World War II, the End of Civilization*. New York: Simon & Schuster, 2008.

Baker, W.J. and J.A. Mangan. *Sport in Africa: Essays in Social History*. New York: Africana Publishing House, 1987.

Balestier, J. "A View of the State of Agriculture in the British Possession in the State of Malacca". *Journal of the Indian Archipelago and Eastern Asia* II (1884): 139–50.

Barlow, Colin. *The Natural Rubber Industry: Its Development, Technology, and Economy in Malaysia*. Kuala Lumpur: Oxford University Press, 1978.

Basalla, George. "The Spread of Western Science". *Science* 156 (1967): 611–22.

———. *The Evolution of Technology*. Cambridge: Cambridge University Press, 1988; reprinted 1990.

Bayly, Susan. "The Evolution of Colonial Cultures: Nineteenth-Century Asia". In *The Oxford History of the British Empire: The Nineteenth Century*, Andrew Porter ed., Chapter 20. Oxford and New York: Oxford University Press, 1999.

Bell, Morag, R. Butlin, and M. Hefferman, eds. *Geography and Imperialism 1820–1940*. Manchester and New York: Manchester University Press, 1995.

Bernstein, William. *A Splendid Exchange: How Trade Shaped the World*. London: Atlantic Books, 2008.

Bilainkin, George. *Hail Penang! Being the Narrative of Comedies and Tragedies in a Tropical Outpost, among Europeans, Chinese, Malays, and Indians*. London: Sampson Low, Marston and Company, 1932.

Blackford, M. *The Rise of Modern Business in Great Britain, the United States and Japan*. Chapel Hill: University of North Carolina Press, 1988.

Bogaars, G.E. "The Tanjong Pagar Dock Company, 1864–1905". *Memoirs of the Raffles Museum* III (1956): 117–243.

Braddell, Roland. *The Lights of Singapore*. London: Muthuen, 1934.

Braddell, T. "Notes on the Chinese in the Straits". *Journal of the Indian Archipelago and Eastern Asia* IX (1855): 109–24.

Brendon, Piers. *The Decline and Fall of the British Empire, 1781–1997*. London: Vintage, 2008.

Breznitz, Dan. *Innovation and the State: Political Choice and Strategies for Growth in Israel, Taiwan and Ireland*. New Haven and London: Yale University Press, 2007.

Brockway, Lucille. *Science and Colonial Expansion: The Role of the British Royal Botanic Gardens*. New York: Academic Press, 1979.

Broeze, Frank, ed. *Private Enterprise, Government and Society: Studies in Western Australian History*. Centre for Western Australian History, University of Western Australia, 1992.

Broeze, Frank, Peter Reeves, and Kenneth McPherson. "Imperial Ports and the Modern Economy: The Case of the Indian Ocean". In *Port and Harbour Engineering,* edited by Adrian Jarvis. Aldershot: Ashgate, 1998.

Brown, Edwin A. *Indiscreet Memories: 1901 Singapore through the Eyes of a Colonial Englishman.* Singapore: Monsoon Books, 2007.

Bryson, Bill. *At Home: A Short History of Private Life.* London: Doubleday, 2010.

Buchanan, Iain. *Singapore in South East Asia: An Economic and Political Appraisal.* London: G. Bell and Sons Ltd, 1972.

Buchanan, R.A. "Institutional Proliferation in the British Engineering Profession, 1847–1914". *Economic History Review* 38, no. 1 (1985): 42–60.

———. "The Diaspora of British Engineering". *Technology and Culture* 27, no. 3 (1986): 501–24.

Buckley, C.B. *An Anecdotal History of Old Times in Singapore.* Singapore, 1902; reprinted, Kuala Lumpur, 1965.

Burkill, H.M. "Murray Ross Henderson, 1899–1983 and some notes on the administration of botanical research in Malaya". *Journal of the Malaysian Branch of the Royal Asiatic Society* 56 (1983): 87–104.

Burroughs, Peter. "Imperial Institutions and the Government of Empire". In *The Oxford History of the British Empire; The Nineteenth Century,* edited by Andrew Porter, Chapter 9. New York and Oxford: Oxford University Press, 1999.

Butcher, J.G. *The British in Malaya, 1880–1941: The Social History of a European Community in Colonial Southeast Asia.* Kuala Lumpur: Oxford University Press, 1979.

Cable, B. *A Hundred-Year History of the P. & O. (Peninsula and Oriental Steam Navigation Company), 1837–1937.* London, 1937.

Cain, P.J. and A.G. Hopkins. *British Imperialism: Innovation and Expansion, 1688–1914.* London, New York: Longman, 1993.

Cain, P.J. and A.G. Hopkins. *British Imperialism: Crisis and Deconstruction, 1914–1990.* London, New York: Longman, 1993.

Cameron, John. *Our Tropical Possession in Malayan India.* London: Smith Wilder, 1865; reprinted, Kuala Lumpur: Oxford University Press, 1965.

Cannadine, D. *Ornamentalism: How the British Empire saw their Empire.* London: Penguin Books, 2002.

———. *Making History Now and Then: Discoveries, Controversies and Explorations.* New York: Palgrave Macmillan, 2008.

Cardwell, D. *The Fontana History of Technology.* London: Fontana Press, 1994.

Carruthers, J.B. "Rubber". In *Twentieth Century Impressions of British Malaya: Its History, People, Commerce, Industries and Resources,* edited by A. Wright

and H.A. Cartwright, pp. 195–96. Singapore, 1908; reprinted, Singapore: Graham Brash, 1989.
Castellani, A. "The Adaptation of European Women and Children to Tropical Climates". *Proceedings of the Royal Society of Medicine* 24 (1931): 95–98.
Chan, Kwok Bun and Claire Chiang, S.N. *Stepping Out: The Making of Chinese Entrepreneurs*. Singapore: Prentice Hall, 1994.
Chan, Wai Kwan. *The Making of Hong Kong Society: The Three Case Studies of Class Formation in Early Hong Kong*. Oxford: Clarendon Press, 1991.
Chance, T. and Williams P. *Lighthouse: The Race to Illuminate the World*. London: New Holland Publishers, 2008.
Chandler, Alfred D. Jr. "Creating Competitive Capability: Innovation and Investment in the United States, Great Britain, and Germany from the 1870s to World War I". In *Favorites of Fortune: Technology, Growth, and Economic Development since the Industrial Revolution,* edited by P. Higonnet, David S. Landes and H. Rosovsky, Chapter 15. Cambridge, Massachusetts: Harvard University Press, 1991.
Chang, Iris. *The Chinese in America: A Narrative History*. New York; Penguin Books, 2003.
Chelliah, D. *A History of the Educational Policy of the Straits Settlements with Recommendations for a New System Based on Vernaculars*. Kuala Lumpur: Government Press, 1947; reprinted, Singapore: Kiat & Co., 1960.
Chew, Ernest and Edwin Lee, eds. *A History of Singapore*. Singapore: Oxford University Press, 1991.
Christopher, P.J. *The British Empire at its Zenith*. London: Croom Helm, 1988.
Clark, P. David. *Germs, Genes and Civilization: How Epidemics Shaped Who We Are Today*. New Jersey: Pearson Education, 2010.
Clark, Gregory. *A Farewell to Alms: A Brief Economic History of the World*. Princeton: Princeton University Press, 2007.
Cowan, Ruth Schwartz. *A Social History of American Technology*. New York and Oxford: Oxford University Press, 1997.
Crawfurd, John. *Journey of an Embassy from the Governor-General to the Courts of Siam and Cochin China*. London: H. Colburn and R. Bentley, 1830; reprinted, Kuala Lumpur, 1967.
Crosby, A.W. *Ecological Imperialism: The Biological Expansion of Europe, 900–1900*. Cambridge: Cambridge University Press, 1986; reprinted 1992.
Davenport-Hines, R.P.T. and Geoffrey Jones. *British Business in Asia since 1860*. Cambridge: Cambridge University Press, 1989.
Delgado, James. *Kamikaze: History's Greatest Naval Disaster*. London: Vintage Books, 2009.
Dharmasena, K. *The Port of Colombo*. Colombo: Ministry of Higher Education, 1980.

Dijkman, Johaness. *Hevea: Thirty Years of Research in the Far East*. Coral Gables: University of Miami Press, 1952.
Dixon, A. "Milk Production in Singapore". *British Malaya*, March 1963, pp. 28–29.
Dobbs, Stephen. *The Singapore River: A Social History 1819–2002*. Singapore: Singapore University Press, 2003.
Doraisamy, T.R., ed. *150 Years of Education in Singapore*. Singapore: TTC Publications Board, Teachers Training College, 1969.
Dower, John. *Japan in War and Peace: Essays on History, Race and Culture*. London: Harper Collins, 1993.
Drabble, John. *Rubber in Malaya, 1876–1922: The Genesis of the Industry*. Kuala Lumpur: Oxford University Press, 1973.
———. "Technology Transfer in Singapore/Malaya during the Colonial Period". *Journal of the Malaysian Branch of the Royal Asiatic Society* 76 (2003): 81–85.
———. "A Note on Agricultural History of Peninsula Malaysia". *Journal of the Malaysian Branch of the Royal Asiatic Society* 82 (2009): 113–17.
Dyos, H.J. and D.H. Aldcroft. *British Transport: An Economic Survey from the Seventeenth Century to the Twentieth*. Harmondsworth: Pelican Books, 1969.
Earl, G.W. "On the culture of Cotton in the Straits Settlements". *Journal of the Indian Archipelago and Eastern Asia* 4 (1850): 720–27.
Edelstein, M. "Foreign Investment and Empire, 1860–1914". In *The Economic History of Britain since 1700*, edited by R. Floud and D. McCloskey. New York: Cambridge University Press, 1994.
Edmond, Rod. *Leprosy and Empire: A Medical and Cultural History*. Cambridge: Cambridge University Press, 2006.
Edwards, Norman. *The Singapore House and Residential Life, 1819-1939*. Singapore: Oxford University Press, 1991.
Emerson, Rupert. *Malaysia: A Study in Direct and Indirect Rule*. New York: Macmillan, 1937; reprinted, Kuala Lumpur: University of Malaya Press, 1964.
Eyal, J. "History as it should be taught". *Straits Times*, 17 July 2010.
Fell, R.T. *Early Maps of South-East Asia*. Oxford and New York: Oxford University Press, 1988.
Ferguson, Niall. *Empire: How Britain Made the Modern World*. London: Penguin Books, 2007.
Ferguson, Niall. *Civilization: The West and the Rest*. London: Allen Lane, 2011.
Fieldhouse, David. *Economics and Empire, 1830–1914*. London: Weidenfeld and Nicholson, 1973.

Finch, James. *Engineering and Western Civilisation*. New York: McGraw Hill, 1951.
Francis, L.H. *The Singapore Chapter of the Narrative of the Expedition of An American Squadron to the China Seas and Japan*. New York: D. Appleton & Company, 1857; reprinted, Singapore: Antiques of the Orient, 1988).
Fraser and Neave Limited. *1883–1983: The Great Years*. Singapore, n.d.
Fraser, Evan D.G. and A. Rimas. *Empires of Food: Feast, Famine, and the Rise and Fall of Civilizations*. New York and London: Free Press, 2010.
Freedman, Maurice. "Immigrants and Associations: Chinese in the 19th Century Singapore". *Comparative Studies in Society and History* III (1960): 25–48.
Freeman, M., R. Pearson and J. Taylor. "Technological change and the governance of joint-stock enterprise in the early nineteenth century: The case of coastal shipping". *Business History* 49, no. 5 (2007): 573–94.
Fremantle, Francis. *A Traveller's Study of Health and Empire*. London: John Ouseley, 1911.
Freese, Barbara. *Coal: A Human History*. London: Arrow Books, 2003.
Garvin, James, L. *The Life of Joseph Chamberlain*. London: Macmillan and Company, 1934.
Glosser, Susan. "Milk for Health, Milk for Profit: Shanghai's Chinese Diary Industry under Japanese Occupation". In *Inventing Nanjing Road: Commercial Culture in Shanghai, 1900–1945,* edited by Sherman Cochran, pp. 207–33. New York: Cornell University, 1999.
Goh Chor Boon. "Imported Technology". *Journal of the Malaysian Branch of the Royal Asiatic Society* 71 (1998): 41–54.
———. *Serving Singapore: A Hundred Years of Cold Storage, 1903–2003*. Singapore: Cold Storage Singapore, 2003.
Green, E.H. "The Political Economy of Empire, 1880–1914". In *The Oxford History of the British Empire; The Nineteenth Century,* edited by Andrew Porter, Chapter 16. New York and Oxford: Oxford University Press, 1999.
Gullick, J.M. *Kuala Lumpur, 1880–1895*. Malaysia: Pelanduk Publications. 1988.
Guillou, M.L. "Technical Education 1850–1914". In *Where Did We Go Wrong? Industry Education and Economy of Victorian Britain*, edited by G. Roderick and M. Stephens, pp. 173–84. Sussex: The Falmer Press, 1981.
Havinden, M. and D. Meredith. *Colonialism and Development: Britain and Its Tropical Colonies, 1850–1960*. London: Routledge, 1993.
Headrick, R. Daniel. *The Tools of Empire: Technology and European Imperialism in the 19th Century*. New York: Oxford University Press, 1981.
———. *The Tentacles of Progress: Technology in the Age of Imperialism, 1850–1940*. New York: Oxford University Press, 1988.

———. *The Invisible Weapon: Telecommunications and International Politics, 1815–1945*. New York: Oxford University Press, 1991.

———. *Power Over Peoples: Technology, Environments, and Western Imperialism, 1400 to Present*. Princeton, New Jersey: Princeton University Press, 2010.

Heussler, Robert. *British Rule in Malaya: The Malayan Civil Service and Its Predecessors 1867–1942*. Oxford: Clio Press, 1981.

Higonnet, P., D. Landes, and H. Rosovsky, eds. *Favourites of Fortune: Technology Growth and Economic Development since the Industrial Revolution*. Cambridge, Massachusetts: Harvard University, 1991.

Hirose, Shin. "Two Classes of British Engineers: An Analysis of Their Education and Training, 1880s–1930s". *Technology and Culture* 51, no. 1 (2010): 388–402.

Hodge, Joseph M. "Science, Development, and Empire: The Colonial Advisory Council on Agriculture and Animal Health, 1929–43". *Journal of Imperial and Commonwealth History* 30, no. 1 (2002): 1–26.

Hough, G.G. "Notes on the Educational Policy of Sir Stamford Raffles". *Journal of the Malayan Branch Royal Asiatic Society* 62 (1969): 155–60.

Howe, Stephen. "Empire and Ideology". In *The British Empire: Themes and Perspectives*, edited by Sarah Stockwell, Chapter 7. Oxford: Blackwell Publishing, 2008.

Hudson, Derek and Kenneth W. Luckhurst. *The Royal Society of Arts, 1754–1954*. London: John Murray, 1954.

Huff, W.G. *The Economic Growth of Singapore: Trade and Development in the Twentieth Century*. Cambridge: Cambridge University Press, 1994.

Hui Po-Keung. "Comprador Politics and Middleman Capitalism". In *Hong Kong's History: State and Society Under Colonial Rule,* edited by Ngo Tak-Wing, Chapter 3. London and New York: Routledge, 1999.

Ingham, B. and C. Simmons. *Development Studies and Colonial Policy*. London: Frank Cass, 1987.

Inkster, Ian. *Science and Technology in History: An Approach to Industrial Development*. London: Macmillan, 1991.

———, ed. *The Steam Intellect Societies: Essays on Culture, Education and Industry, circa 1820–1914*. University of Nottingham, Department of Adult Education, 1985.

———. "The Trouble with Technology: Comments on the Experience of Singapore under Entrepot Colonialism". *Journal of the Malaysian Branch of the Royal Asiatic Society* 73 (2000): 107–15.

——— and Fumihiko Satofuka, eds. *Culture and Technology in Modern Japan*. London: I. B. Tauris, 2000.

Jackson, J.C. *Planters and Speculators: Chinese and European Agricultural Enterprises in Malaya 1786–1921*. Kuala Lumpur: University of Malaya Press, 1968.

Jackson, R.A. *Immigrant Labour and the Development of Malaya*. Federation of Malaya: Government Press, 1961.
Jacques, Martin. *When China Rules the World*. London: Allen Lane, 2009.
Jarvis, Adrian, ed. *Port and Harbour Engineering*. Aldershot: Ashgate, 1998.
Jennings, Eric. *Mansfield: Transport and Distribution in South-East Asia*. Singapore: Meridian Communications, 1973.
Johnson, Robert. *British Imperialism*. New York: Palgrave Macmillan, 2003.
Joseph, K.T. "Agricultural History of Peninsula Malaysia: Contributions from Indonesia". *Journal of the Malaysian Branch of the Royal Asiatic Society* 81, Part 1 (2008): 7–18.
Kaur, Amarjit. *Bridge and Barrier: Transport and Communications in Colonial Malaya, 1870–1957*. Singapore: Oxford University Press, 1985.
Kemp, Tom. *Industrialisation in the Non-Western World*. London: Longman, 1983.
Kennedy, P.M. "Imperial Cable Communications and Strategy, 1879–1914". *English Historical Review* 86 (1971): 728–52.
———. *The Rise and Fall of Great Powers: Economic Change and Military Conflict from 1500 to 2000*. London: Fontana Press, 1988.
———. *The Rise and Fall of British Naval Mastery*. London: Fontana Press, 1991.
Kesner, R.M. *Economic Control and Colonial Development: Crown Colony Financial Management in the Age of Joseph Chamberlain*. Westport, Conn.: Greenwood Press, 1981.
Khoo, Salma Nasution. *Kinta Valley: Pioneering Malaysia's Modern Development*. Perak, Malaysia: Perak Academy, 2005.
Khoo Teng Soon. *Interview*, Oral History Unit. Singapore Archives, 11 September 1984.
Kidd, Benjamin. *Social Evolution*. New York and London: Macmillan and Co., 1894.
Kratoska, Paul. *Honourable Intentions: Talks on the British Empire in South-East Asia Delivered at the Royal Colonial Institute, 1874–1928*. Singapore: Oxford University Press, 1983.
Kubicek, R.V. *The Administration of Imperialism: Joseph Chamberlain at the Colonial Office*. Durham: Duke University Commonwealth Studies Centre, 1969.
Kubicek, Robert. "British Expansion, Empire, and Technological Change". In *The Oxford History of the British Empire; The Nineteenth Century*, edited by Andrew Porter, Chapter 12. New York and Oxford: Oxford University Press, 1999.
Kwa Chong Guan. *Singapore, a 700-year History: From Early Emporium to World City*. Singapore: National Archives of Singapore, 2009.

Landes, David S. *The Unbound Prometheus: Technological Change and Industrial Development in Western Europe from 1750 to the Present*. Cambridge: Cambridge University Press, 1969.

———. *The Wealth and Poverty of Nations*. London: Little, Brown and Company, 1998.

Laxon, W.A. *The Straits Steamship Fleets*. Sarawak: The Sarawak Steamship Company Berhad, 2004.

Lee, Edwin. "The Colonial Legacy". In *Management of Success: The Moulding of Modern Singapore*, edited by K.S. Sandhu and Paul Wheatley, Chapter 1. Singapore: Institute of Southeast Asian Studies, 1989.

———. *Historic Buildings of Singapore*. Singapore: Preservation of Monuments Board, 1990.

———. *The British as Rulers: Governing Multicultural Singapore, 1867–1914*. Singapore: Singapore University Press, 1991.

Lee Kuan Yew. *From Third World to First: The Singapore Story, 1965–2000*. Singapore: Times Editions, 2000.

Lee Lai To, ed. *Early Chinese Immigrant Societies: Case Studies from North America and British Southeast Asia*. Singapore: Heinemann Asia, 1988.

Lee Poh Ping. *Chinese Society in Nineteenth Century Singapore*. Singapore: Oxford University Press, 1978.

Lee Yong Kiat. *The Medical History of Early Singapore*. Tokyo: Southeast Asian Medical Information Center, 1978.

Liew Kai Khiun. "Planters, Estate Health & Malaria in British Malaya (1900–1940)". *Journal of the Malaysian Branch of the Royal Asiatic Society* 83, Part 1 (2010): 91–115.

Lim Chong Yah. *Economic Development of Modern Malaya*. Kuala Lumpur: Oxford University Press, 1979.

Lockhart, B. *Return to Malaya*. London: Putnam, 1936.

Loh Kah Seng. *Making and Unmaking the Asylum: Leprosy and Modernity in Singapore and Malaysia*. Petaling Jaya: Strategic Information and Research Development Centre, 2009.

Maaruf, S. *Malay Ideas on Development: From Feudal Lord to Capitalist*. Singapore: Times Books, 1988.

MacFarlane, A. and I. MacFarlane. *The Empire of Tea*. Woodstock and New York: The Overlook Press, 2009.

MacLeod, Roy and Deepak Kumar, eds. *Technology and the Raj: Western Technology and Technical Transfers to India 1700–1947*. New Delhi and London: Sage, 1995.

Makepeace Walter, Gilbert E. Brooke and Roland St. J. Braddell, eds. *One Hundred Years of Singapore*. London: John Murray, 1921; reprinted, Singapore: Oxford University Press, 1991.

Manderson, Lenore. *Sickness and the State: Health and Illness in Colonial Malaya, 1870–1940*. Melbourne: Cambridge University Press, 1996.

———. "Blame, Responsibility and Remedial Action: Death, Disease and the Infant in Early Twentieth Century Malaya". In *Death and Disease in Southeast Asia*, edited by Norman Owen, Chapter 12. Singapore: Oxford University Press, 1987.

Mangan, J.A. "Ethics and Ethnocentricty: Imperial education in British Tropical Africa". In *Sport in Africa: Essays in Social History*, edited by W.J. Baker and J.A. Mangan, Chapter 6. New York: Africana Publishing House, 1987.

———, ed. *The Imperial Currciculum: Racial Images and Education in the British Colonial Experience*. New York, London: Routledge, 1993.

Marriner, Sheila. *The Economic and Social Development of Merseyside*. London: Croom Helm, 1982.

Marshall, Adrian G. *The Singapore Letters of Benjamin Cook 1854–1855*. Singapore: Landmark Books, 2004.

Marshall, Peter, ed. *The Cambridge Illustrated History of the British Empire*. Cambridge: Cambridge University Press, 1996.

Marshall, Peter J. "Empire Authority in the Later Eighteenth Century". *Journal of Imperial and Commonwealth History* 15, no. 2 (1987): 105–22.

McNab, Colin and Robert Mackenzie. *From Waterloo to the Great Exhibition: Britain 1815–1852*. Oliver & Boyd, 1982.

Melosi, Martin, V. *Garbage in the Cities: Refuse, Reform and the Environment*. Pittsburgh: University of Pittsburgh Press, 2005.

———. *The Sanitary City: Environmental Services in Urban America from Colonial Times to the Present*. Pittsburg: University of Pittsburg Press, 2008.

Meredith, David. "The British Government and Colonial Economic Policy 1919–1939". *Economic History Review* 28 (1975): 484–99.

Metcalfe, Thomas R. "Imperial Towns and Cities". In *The Cambridge Illustrated History of the British Empire*, edited by Peter Marshall, Chapter 8. Cambridge: Cambridge University Press, 1996.

Milton, Giles. *Nathaniel's Nutmeg: How One Man's Courage Changed the Course of History*. London: Sceptre, 1999.

Minchinton, W.E. "British Ports of Call in the Nineteenth Century". *The Mariner's Mirror* 62, no. 2 (1976): 145–57.

Miners, Norman. "Industrial Development in the Colonial Empire and the Imperial Economic Conference at Ottawa 1932". *Journal of Imperial and Commonwealth History* 30, no. 2 (2002): 53–76.

Mintz, Sidney. *Sweetness and Power: The Place of Sugar in Modern History*. New York: Penguin, 1985.

Mohanram, Radhika. *Imperial White: Race, Diaspora and the British Empire*. Minneapolis/London: University of Minnesota Press, 2007.

Montgomerie, W. "Report upon the present state of the Honourable Company's Botanical Garden at Singapore, 1ˢᵗ February 1827" *Journal of the Malaysian Branch of the Royal Asiatic Society* XLII (1969): 62–65.
Moore, Donald and J. Moore. *The First 150 Years of Singapore*. Singapore: Donald Moore Press, 1969.
Mokyr, Joel. *The Lever of Riches: Technological Creativity and Economic Progress*. New York and Oxford: Oxford University Press, 1990.
Morris, Ian. *Why the West Rules — For Now: The Patterns of History and What They Reveal about the Future*. London: Profile Books, 2010.
Morris, Jan. *Pax Britannica: The Climax of an Empire*. Penguin Books, 1981.
Ngo Tak-Wing, ed. *Hong Kong's History: State and Society under Colonial Rule*. London and New York: Routledge, 1999.
Odagiri, Hiroyuki and Akira Goto. *Technology and Industrial Development in Japan: Building Capabilities by Learning, Innovation and Public Policy*. Oxford: Clarendon Press, 1996.
Ong Chit Chung. *Operation Matador: Britain's War Plans against the Japanese 1918–1941*. Singapore: Times Academic Press, 1997.
Owen, Norman, ed. *Death and Disease in Southeast Asia*. Singapore: Oxford University Press, 1987.
Pacey, Arnold. *The Maze of Ingenuity: Ideas and idealism in the Development of Technology*. Cambridge, Massachusetts: The MIT Press, 1992.
Parkison, Northcote C. *Britain in the Far East: The Singapore Naval Base*. Singapore: Donald Moore, 1955.
Paterson, Michael. *Voices From Dickens' London*. Cincinnati: David and Charles, 2006.
Petersen, Christian. *Bread and the British Economy, 1770–1870*. Aldershot, England: Scolar Press, 1995.
Pfammatter, Ulrich. *The Making of the Modern Architect and Engineer: The Origins and Development of a Scientific and Industrially Oriented Education*. Berlin: Birkhauser, 2000.
Platt, C.D.M. "Economic Factors in British Policy during the 'New Imperialism'". *Past and Present* 39 (1968): 120–38.
Porter, Andrew. *European Imperialism, 1860–1914*. London: Macmillan, 1994.
———, ed. *The Oxford History of the British Empire: The Nineteenth Century*. Oxford and New York: Oxford University Press, 1999.
Porter, Bernard. *The Absent-Minded Imperialists: What the British Really Thought about Empire*. Oxford: Oxford University Press, 2004.
Rawlinson, J. "Sanitary Engineering: Sanitation". In *A History of Technology*, edited by Charles Singer, Chapter 4. Oxford: Oxford University Press, 1958.
Read, Donald. *Cobden and Bright: A Victorian Political Partnership*. London: Edward Arnold, 1967.

Reid, Anthony. *Southeast Asia in the Age of Commerce, 1450–1680*. New Haven and London: Yale University Press, 1988.
Ridley, H.N. "Botany". In *Twentieth Century Impressions of British Malaya: Its History, People, Commerce, Industries and Resources*, edited by A. Wright and H.A. Cartwright, pp. 185–90. Singapore, 1908; reprinted, Singapore: Graham Brash, 1989.
Regnier, Phillipe. *Singapore: City State in South East Asia*. Honolulu: University of Hawai'i Press, 1987.
Robbins, Keith. *The Eclipse of a Great Power: Great Britain, 1870–1975*. London: Longman, 1983.
Robertson, P.L. "Technical Education in the British Shipbuilding and Marine Engineering Industries 1863–1914". *Economic History Review* 27, no. 2 (1974): 222–35.
Roderick, G. and M. Stephens. *Where Did We Go Wrong? Industry Education and Economy of Victorian Britain*. London: The Falmer Press, 1981.
Rosecrance, Richard. *The Rise of the Trading State: Commerce and Conquest in the Modern World*. New York: Basic Books, 1986.
Rudner, M. "Colonial Education Policy". In *Development Studies and Colonial Policy*, edited by B. Ingham and C. Simmons, Chapter 8. London: Frank Cass, 1987.
Ryan, Jan. "The Business of Chinese Immigration". In *Private Enterprise, Government and Society: Studies in Western Australian History*, edited by F. Broeze. Centre for Western Australian History, University of Western Australia, 1992.
Saham, J. *British Industrial Investment in Malaysia 1963–1971*. Kuala Lumpur: Oxford University Press, 1980.
Savage, Victor. *Western Impression of Nature and Landscape in Southeast Asia*. Singapore: Singapore University Press, 1984.
Seah Eu Chin. "The Chinese in Singapore: General Sketch of the Numbers, Tribes and Avocations of the Chinese in Singapore". *Journal of the Indian Archipelago* II (1848): 283–89.
Semmel, Bernard. *The Rise of Free Trade Imperialism*. Cambridge: Cambridge University Press, 1970.
———. *Liberalism and Naval Strategy: Ideology, Interest and Sea Power during the Pax Britainnica*. Boston: Allen & Unwin. 1986.
Shellabear, W.G. "Baba Malay: An Introduction to the Language of the Straits-born Chinese". *Journal of the Royal Asiatic Society* 65 (1913): 49–63.
Shennan, Margaret. *Out in the Midday Sun: The British in Malaya, 1880–1960*. London: John Murray, 2000.
Sidney, J.R.H. *Malay Land*. London: Cecil Palmer, 1926.
Silcock, Thomas H., ed. *Readings in Malayan Economics*. Singapore: Eastern Universities Press, 1961.

Sim, Katherine. *Malayan Landscape*. Singapore: Asia Pacific Press, 1969.
Singer, Charles, ed. *A History of Technology*. Oxford: Oxford University Press, 1958.
Song Ong Siang. *One Hundred Years' History of the Chinese in Singapore*. London: John Murray; reprinted, Singapore: Oxford University Press, 1991.
Standage, Tom. *An Edible History of Humanity*. New York: Walker & Company, 2009.
Stanworth, J. and C. Stanworth. *Work 2000: The Future for Industry, Employment and Society*. London: Paul Chapman, 1991.
Staples, A.F. "Malaya Revisited". *British Malaya*, October 1935, pp. 137–39.
Stearns, Peter N. *The Industrial Revolution in World History*. Boulder: Westview Press, 1993.
Stockwell, Sarah, ed. *The British Empire: Themes and Perspectives*. Oxford: Blackwell Publishing, 2008.
Suzuki, Shogo. *Civilization and Empire: China and Japan's Encounter with European International Society*. London and New York: Routledge, 2009.
Swettenham, Frank. *British Malaya*. London: G. Allen & Unwin, 1955.
Tagliacozzo, E. "The Lit Archipelago: Coast Lighting and the Imperial Optic in Insular Southeast Asia, 1800–1910". *Technology and Culture* 46, no. 2 (2005): 306–28.
Tan Ee Leong. "The Chinese Banks Incorporated in Singapore and Malaya". In *Readings in Malayan Economics,* edited by Thomas S. Silcock. Singapore: Eastern Universities Press, 1961.
Tate, D.J.M. *Straits Affairs: The Malay World and Singapore*. Hong Kong: John Nicholson, 1989.
Tatsuno, Sheridian M. *Created in Japan: From Imitators to World-Class Innovators*. New York: Harper & Row, 1990.
Telecommunication Authority of Singapore. *A Hundred Years of Dedicated Telephone Service in Singapore*. Singapore: International Press, 1979.
Teo Siew Eng and Victor Savage. "Singapore Landscape: A Historical Overview of Housing Image". In *A History of Singapore*, edited by Ernest Chew and Edwin Lee, Chapter 14. Singapore: Oxford University Press, 1991.
Tessa, Morris-Suzuki. *The Technological Transformation of Japan: From the Seventeenth to the Twenty-first Century*. Cambridge: Cambridge University Press, 1994.
The Singapore Chapter of Zieke Reiziger; or Rambles in Java and the Straits in 1852 by a Bengal Civilian. London: Simpkin, Marshall and Co.; republished Singapore: Antiques of the Orient, 1984.
Thomas, Antony. *Rhodes*. London: BBC Books, 1996.
Thomson, J.T. "Account of the Horsburg Lighthouse". *Journal of the Indian Archipelago and Eastern Asia* 6 (1853): 376–498.

———. *Some Glimpses into Life in the Far East*. London: Richardson & Co, 1965.
Tomlinson, B.R. "Economics and Empire: The Periphery and the Imperial Economy". In *The Oxford History of the British Empire*, edited by Andrew Porter, Chapter 3. Oxford and New York: Oxford University Press, 1999.
Tregonning, K.G. *Straits Tin: A Brief Account of the First Seventy-Five Years of the Straits Trading Company Limited, 1887–1962*. Singapore: The Straits Times Press, n.d.
———. *The Singapore Cold Storage Company, 1903–1966*. Singapore: The Straits Times Press, 1966.
———. "Tertiary Education in Malaya: Policy and Practice 1905–1962". *Journal of the Malayan Branch of the Royal Asiatic Society* 63, Part 1 (1990): 1–14.
Trocki, Carl A. *Opium and Empire: Chinese Society in Colonial Singapore, 1800–1910*. Ithaca and London: Cornell University Press, 1990.
———. *Prince of Pirates*. Singapore: Singapore University Press, 1979.
Tsang, Steve. *A Modern History of Hong Kong*. London and New York: I. B. Tauris, 2004.
Turnbull, C.M. "The European Mercantile Community in Singapore 1819–1867". *Journal of South East Asia History* 10, no. 1 (1969): 12–35.
———. *The Straits Settlements, 1826–1867: Indian Presidency to Crown Colony*. London: The Athlone Press, 1972.
———. *A History of Singapore, 1819–1988*. Singapore: Oxford University Press, 1989.
Twells, Alison. *The Civilising Mission and the English Class, 1792–1850*. New York: Palgrave Macmillan, 2009)
Vogel, Ezra. *The Four Little Dragons: The Spread of Industrialization in East Asia*. Cambridge: Harvard University Press, 1991.
Walton, Oliver C. "Officers or Engineers? The Integration and Status of Engineers in the Royal Navy, 1847–60". *Institute of Historical Research* 77, no. 196 (2004): 178–201.
Wang Gungwu. *China and the Chinese Overseas*. Singapore: Times Academic Press, 1991.
Warren, James F. *Rickshaw Coolie: A People's History of Singapore 1880–1940*. Singapore: Singapore University Press, 2003.
Wasserstrom, Jeffrey, N. *Global Shanghai, 1850–2010: A History in Fragments*. London and New York: Routledge, 2009.
Watson, K. "Rulers and Ruled: Racial Perceptions, Curriculum and Schooling in Colonial Malaya and Singapore". In *The Imperial Currciculum: Racial Images and Education in the British Colonial Experience*, edited by J.A. Mangan, Chapter 10. New York, London: Routledge, 1993.

Watts, Sheldon. *Epidemics and History: Disease, Power and Imperialism*. New Haven and London: Yale University Press, 1997.

———. *Disease and Medicine in World History*. New York and London: Routledge, 2003.

Webster, Anthony. *The Twilight of the East India Company: The Evolution of Anglo-Asian Commerce and Politics 1790–1860*. Woodbridge and Rochester: Boydell Press, 2009.

Weiss, Eugene H. *Chrysler, Ford, Durant and Sloan: Founding Giants of the American Automotive Industry*. Jefferson, N. C.: McFarland, 2003.

Wilkes, Charles. *The Singapore Chapter of the Narrative of the United States Exploring Expedition*. Philadelphia: Lea & Branchard, 1845; reprinted, Singapore: Antiques of the Orient, 1984.

Williams, Michael. "The Relations of Environmental History and Historical Geography". *Journal of Historical Geography* 20, no. 1 (1994): 3–21.

Winstedt, R.O. "Singapore, Past and Present". *British Malaya* 3 (1938): 290–91.

———. "The Land I Love". *British Malaya* 9 (1940): 71–72.

Winstedt, R.O. *A History of Malaya*. Kuala Lumpur and Singapore: Marican & Sons, 1982.

Woodine, Philip. *The Industrial Revolution*. University of Huddersfield: Huddersfield Pamphlets in History and Politics, 1993.

Wong, F. and H.Y. Gwee. *Official Reports on Education: Straits Settlements and the Federated Malay States 1870–1939*. Singapore: Pan Pacific Books, 1983.

Wong Lin Ken. "The Trade of Singapore, 1819–1869". *Journal of Malaysian Branch Royal Asiatic Society* 30, Part 4 (1960): 5–315.

———. "Singapore: Its Growth as an Entrepot Port, 1819-1914". *Journal of Southeast Asian Studies* 9, no. 1 (1978): 50–84.

Wong Yee Tuan. "More Than a Tea Planter: Jon Archibald Russell and His Businesses in Malaya, 1899-1933". *Journal of the Malaysian Branch of the Royal Asiatic Society* 83, Part 1 (2010): 29–51.

Wright, A. and H.A. Cartwright, eds. *Twentieth Century Impressions of British Malaya: Its History, People, Commerce, Industries and Resources*. London, 1908; reprinted, Singapore: Graham Brash, 1989.

Wright, R.A. *Annual Report on the Veterinary Department*. Singapore: Government Printing Office, 1949.

Wurtzburg, C.E. "Higher Education in Malaya". *British Malaya*, February 1940.

———. "Singapore — Past and Present". *British Malaya*, December 1940.

Yap Pheng Geck. *Scholar, Banker, Gentleman Soldier: The Reminiscences of Dr Yap Pheng Geck*. Singapore: Times Books, 1982.

Yen Ching Hwang. *Class Structure and Social Mobility in the Chinese Community in Singapore and Malaya 1800–1911*. University of Adelaide, Department of History, Working Paper No. 15, 1983.

———. *A Social History of the Chinese in Singapore and Malaya, 1800–1911*. Singapore: Oxford University Press, 1986.

Yeoh, Brenda S.A. *Contesting Space: Power Relations and the Urban Built Environment in Colonial Singapore*. Singapore: Oxford University Press, 1996.

Yip Yat Hoong. *The Development of the Tin Mining Industry of Malaya*. Kuala Lumpur: University of Malaya Press, 1969.

Yong Ching Fatt. *Tah Kah Kee: The Making of an Overseas Chinese Legend*. Singapore: Oxford University Press, 1987.

Young, G. M. *Victorian England: Portrait of an Age*. Oxford: Oxford University Press, 1964.

Index

A
Adas, Michael, 23–25
Aeria, D.A., 201
Africa, partition of, 14
"Age of Empire", 14
agency houses, 152, 231, 236
    role of, 35–42
agents of change, 21
Agri-Horticultural Society, 146, 148
Agricultural Bulletin of the Malay Peninsula, The, 150
Agricultural Bulletin of the Straits and Federated Malay States, The, 150
agricultural commodities, 143
agricultural science, 159–60
"agricultural revolution", 159
agriculturalists, ventures of, 152–59
Aitken, J., 201
air-conditioning, 109, 112
Aird, John, 76
Albert Dock, 70–71
Albert Dock Hospital, 129
Amery, Leopold, 161
Anderson Bridge, 113
Anderson, John, 75–76
Anglo-American Telegraph Company, 95

"Anglobalization", 14, 18
Anglo-Chinese War, 38
Anglo-Dutch Treaty, 145
Angus, C.S., 201
Annual Report of the Straits Settlements, 111, 211
Annual Report on Education, 203
Archdeacon, S.B., 223
Architectural Association, 48
Arkwright, Richard, 13
"articulate classes", 26, 212
Aspdin, Joseph, 118
Association of Engineers, 50
Association of Teachers in Technical Institutes, 216
Atlantic cable, 95
Australian cattle meat, 171, 190
Austrian Lloyd's, 74

B
Baber, Zaheer, 27, 161
bacteriology revolution, 122
Balestier, James, 153
Balestier Plantation, 153–54
banking businesses, in Singapore, 80
Banks, Joseph, 64
Banque de Indo-Chine, 80

barbarism, 14
Basalla, George, 162, 164
battery system, 101
Battle of Shangani River, 16
Baxter, James, 49
Bazalgette, Joseph, 136
Behn Meyer, 39
Bell, Alexander Graham, 100
Bell Brothers Limited, 56
Bell & Edison, 101
Bell Rock, 46
Ben Line, 79
Bengal Medical Board, 147
Bengal Medical Corp, 145
Bentong, vessel, 50
Berkshire boars, 180
Bibbly Line, passenger ship, 36
bicycle, invention of, 103, 116
Bidwell, R.A. John, 48
Billiton, ship, 79
Bintang, steamer, 49
Bird, Isabella, 106
Birkenhead Docks, 70
Black, Kenneth, 137
"Black Ships", 72
Blue Funnel, passenger ship, 36, 79
"blue-water school", 82
boat-dwellers, 3
Boat Quay, 68
Bogaardt, Theodore Cornelius, 79
Boilat, Abbe, 25
Borneo Company Limited, 37, 80
Borneo Wharf, 172, 174, 181–82, 188
Botanical and Experimental Garden, 145
Botanic Gardens, Calcutta, 144–45
Botanic Gardens, Ceylon, 151
Botanic Gardens, Mauritius, 149
Botanic Gardens of Singapore, 8, 148–50, 152
 establishing the, 145–47

Boustead, Edward, 36, 38, 57
"boy", head servant, 174, 182
Braddell, Roland, 183
Braddell, Thomas, 39
Braganza, ship, 78
bread, quality of, 176
Britain
 bankrupt nation, 244
 economic interest in Singapore, 3
 education in, 219
 engineers in, 43
 former colonies, 28
 government telegraph operations, 98
 leading colonialist, as, 18–23
 naval supremacy, 65
 raw materials, and, 21
 shipbuilding industry, 66, 215–16
 shipping companies, 77–78
 world trade, and, 81
British Army, 56
British Australian Telegraph Company Limited, 97
British banks, 232–33
British Empire, 3, 5–7, 14–15, 17, 19, 22, 25–26, 28–29, 32, 35, 52–54, 64–66, 71, 73, 80, 82, 85, 94, 95, 101, 107, 129, 137, 143, 153, 200, 211, 215, 221, 226, 241
 cable network in, 99
 leper colonies in, 135
 role of the port in, 81–83
 territorial extent, 18
British engineers, 65
 arrival of, 42–48
 contributions of, 49–53
 Singapore, in, 44, 46
British India, 44, 52, 95, 136, 196, 201, 212

Index

British India Steam Navigation Company, 78
British law, 20
British-made goods, 21
British Malaya, 134, 211, 215
British officials, and contact with locals, 55
British Queen, vessel, 87
British Royal Navy, 43, 59, 66, 71
British superiority, 25–26, 53–57
  see also Western superiority
Brown, Edwin, 104
Buchanan, R.A., 43, 56
Bun Hin, 230
Burkill, H.M., 152
Burstall, H., 108
Burton, Richard Henry, 60
Bushell, George, 146
business associates (*guanxi*), 228

C
C. Dupire & Co., 111
Cable & Wireless, 115
Cadogan, Alexander, 86
Cain, Peter, 15
Calcutta Agricultural Society, 153
Calcutta Botanic Gardens, 144–45
Cambridge School of Agriculture, 161
Cameron, John, 36, 57, 130, 154, 169
Campbell, company, 169
Campbell, J. Argyle, 210
Cannadine, David, 53, 240
canned fish, 158
Canning, Lord, 97
canons, invention of, 30
Cantley, Nathaniel, 149–50
Cape L'Aghulas, lighthouse at, 45
capitalism, 20, 228
Cardwell, Daniel, 116
Cardwell, Donald, 94

Carey, William, 25
Cargill, T.C., 123
Carlyle, Thomas, 20
Carnation, 169
"Carrier" air-conditioner, 109
Carrington, W.T., 123
Carroll, John, 223
Carruthers, J.B., 151
Catholic Order, 197
cattle meat, Australian, 171, 190
Cavenagh Bridge, 113
Cavenagh, William Orfeur, 113
Celtic, vessel, 87
cement
  "Condor", 111
  "Portland", 45, 111, 118
Central Engine Works, 50
Central Pacific Corporation, 89
Ceylon, Botanic Gardens, 151
Chadwick, Edwin, 136
Chamberlain, Joseph, 22–23, 26, 129, 139, 221
Chandler, Alfred, 227
Chang, Iris, 58
Charles II, King, 156
Chartered Bank, 232
Chartered Bank of India, Australia and China, 80
"Chartism", 20
chauvinism, 5, 244
Cheang Hong Lim, 229
chettiars (moneylenders), 232
Chew Nam Seng, 189
Chicago meat-packing district, 181
chilled meat, 171, 190
China Company of Liverpool, 74
Chin Bee, 189
Chinese agriculturalists, ventures of, 152–59
Chinese banking, 10
  growth of, 231–34

Chinese businesses, 222–24, 227–31
Chinese comprador, 7, 91, 192, 231, 232–33
   role of, 35–42
Chinese coolies, 39, 41, 73–74, 89–90
   see also Indian coolies
Chinese diaspora, 38, 58
Chinese education, 200, 210, 212
Chinese elites, 242
Chinese immigrants
   Singapore, in, 38–39
   United States, in the, 62, 235
Chinese middlemen, 7, 39, 230
Chinese miners, 224
Chinese nationalism, 210
Chinese shipbuilding, 86
cholera, 132–33, 137, 140–41, 178
Christianity, 24–25, 28, 33
Chulalongkorn University, 218
Church of England, 134
Churchill, Winston, 86
Chusan, ship, 78
City Hall, 113
civil engineering, 42, 110–14, 120
Civil War, 168
"civilization", definition, 19
"civilizing mission", ideology, 1, 3, 7, 18, 23–26, 57, 240
Clark, Andrew, 22, 99, 100
Clark, Gregory, 28, 30, 34
Cleaveland, Norman, 112, 183
coal, loading of, 72–73, 89
coaling station, 71–72, 81
coffee industry, 151
Cold Storage
   see Singapore Cold Storage
Cold Storage Creameries Ltd, 182
"Cold Storage Danger", 173
Coleman, George Drumgoole, 44, 48
"collaborators of colonialism", 16

Collyer, Captain, 42
Colomb, J.C.R., 82–83
colonial architecture, 112
colonial gardens, 144
"colonial modernity", 24
Colonial Office, 23, 99, 100
colonial politics, 3
colonial ports, 7
colonial revenues, 23, 32
"colonial style", 245
Colonial Stocks Act, 23
colonialism, 5–6, 10, 18, 24, 121, 161, 227, 230, 241–44
"colour bar", issue of, 53–54
communicable diseases, 120, 132
comparative advantage, law of, 21
comprador, Chinese, 7, 91, 192, 231, 232–33
   role of, 35–42
condensed milk, 178–79
"Condor" cement, 111
   see also "Portland cement", 45, 111, 118
Confucian order, 231
Conrad, Joseph, 64
"constructive imperialism", 22
convict labour, 41–42, 69, 113, 136
Coode, J., 110
Cook, Benjamin, 46
Cook, James, 15, 64–65
"Coolie Trade", 39, 58
Cooperative Milk Condensery Friesland, 178
corporate culture, 233
Corporation of the Seamen's Hospital Society, 129
corruption, 241
Crawfurd, John, 36, 65, 157, 160
"cream cracker", 169
Crimean War, 94, 114
Crompton, Samuel, 13

Index

Crosby, Alfred, 131, 140, 159
Crown Colony, 19, 21, 23
Crystal Palace, 13, 29
cultural contacts, 17
"cultural diffusion", 17
cultural legacy, 233
cultural perceptions, 212
Culture System, 37, 144
Cumberbatch and Company, 192
Cumming, John, 25
Cummings, Constance Gordon, 73
Cuynyngham-Brown, Sjovald, 52
Cycle & Carriage, 105

D

Daily Press, 101
Daily Times, 126
Dalhousie, Lord, 94
d'Almeida, Jose, 153, 160
Dampier, William, 65
"Dark Continent", 33
Davis, John, 241
decolonization, 23
"Deep Sea Cables, The", 95
Denny, N.B., 227
Diamond, warship, 85
Dijkman, Johaness, 160
Dixon, Alec, 178–79
Douglas, John, 47
"Downward Filtration" theory, 201
"Dragon's Teeth, The", 38
Drake, Francis, 65
dredging machinery, 69
"drug foods", 153, 165
Drysdale, J.H., 51
Du Port, William J., 70
Duke of Wellington, 31
Duncan, warship, 85
Dutch East Indies, 37, 144, 162
duty-free ports, 21
Dyer, Thiselton, 149

E

Earle, T.E., 223
East India Company, 2, 11, 44, 64, 94, 96, 136, 145, 201
Eastern Extension Australasia and China Telegraph Company (E.E.A. & C.), 97, 100, 115
Eastern Telecommunications Philippine Inc., 115
Eastern Telegraph Company, 97, 99
Ecole Centrale des Arts, 215
Ecole des Mines, 215
Ecole Polytechnique, 215
"economic botany", 143
Economic Gardens, 149
Edinburgh Bridge, 113
Edison, Thomas Alva, 35, 107
Education Act, 215
education change, 197–98
    obstacles to, 210–17
Education Code, 202
education expenditure, 200
educational reforms, 198–203
Edward Boustead & Company, 37, 80
Egypt
    irrigation system, 76
    railways, 87
electric system, 107–109, 169
Elgin iron bridge, 47, 113
Elizabeth, vessel, 49
Elliot, Charles, 160
Emden, ship, 98
Emerson, Rupert, 54
Empire Dock, 77, 188
Empire Hotel, 192
Endeavour, ship, 15
"engineering imperialism", 43
engineering schools, 215
Engledow, Frank Leonard, 161
English education, 196, 199–203, 208, 212–16

English-medium school, 196, 199
English Printing Office, 203–204
Esplanade, 110
"estate hospitals", 133–34
ethnic groups, in Singapore, 122
Eureka Motor, 56
European agency house, 231, 236
   role of, 35–42
European agriculturalists, ventures of, 152–59
European community, in Singapore, 182–84, 187
European firms, in Singapore, 36
European hegemony, 24
Experimental Gardens, 150

F

Faber, Major, 46–47
Faraday, Michael, 147
"Father of Tropical Medicine", 130
Federated Malay States, 133, 139, 152, 174, 199, 211, 213, 227
   expenditure on education, 200
Ferguson, Niall, 14, 28
Fieldhouse, David, 14
filter system, of water, 127, 137
filth theory, 122
"firepower revolution", 16
"First Pharos of the Eastern Seas", 46
First World War, 18, 23, 48, 98, 105, 156, 181, 202, 216
fisheries, 157–58
Fist of Fury, The, film, 63
food hygiene, 131
food processing, 168
food technology, 9, 168
Ford, Henry, 117
Formosa, ship, 78
Fort Canning, 42, 47, 145
Fortnightly, 36
Fowler, Gilbert, 127

Franklin, Captain, 65, 160
Fraser & Chalmers Engineering Works, 77
Fraser and Neave, 51, 57, 192
Fraser, John, 51
Freese Barbara, 71
free trade, 21, 59, 243
   abolition of, 225
French Messageries Maritimes, 74
French schools, 215
Friesian cattle, 179, 193
Froggart, Leslie, 183
frozen meat, 171, 190
"fruit of kings", 156
future trading, 102

G

gambier, cultivation of, 154–56
"Gambier King", 165
gas lighting system, 107–109
Gas, Water and Electricity Departments, 55
Gatling gun, 16
"gentlemanly capitalists", 15
"geographic relocation" process, 17
geopolitics, 32, 82–83
General Engineering Works, 56
germ theory, 120, 122
Glen Kine, passenger ship, 36
Glenogle, steamer, 72
"Gold Mountains" of California, 58
"Gold Rush", 235
"Golden Chersonese, the", 81
Goode, Bill, 183
Goodman, Lester, 109
Goodwood Hotel, 48
Government Hill, 42, 47
Government House, 108
Government Technical College, 206
Graphic, The, 74
Great Britain, *see* Britain

Index

Great Depression, 77, 175, 180
"Great Exhibition, The", 13–14
greenhouse, heated, 157
grocery trade, 168, 175
Grove Road Bridge, 48
guanxi (business associates), 228
Guillemard, Laurence, 213
Gulland, W.G., 40
"gunboat diplomacy", 2
gunpowder, 30
Guthrie, Alexander, 36, 37, 57
Guthrie & Company Limited, 37, 80
gutta percha, 94, 96, 147, 149
Gutta Percha Company of London, 147

H
Hancock, Thomas, 147
Hargreaves, James, 13
Hargreaves, William, 49
Harper, Gilfillan & Company, 37, 80
Hartley, Jesse, 70
Hawks, Francis, 70
Headrick, Daniel, 15, 17, 31, 52, 221–22
healthy living, promotion of, 8
Heenan and Froude Incinerators, 124
Heinz, 169
Henry Waugh & Company, 37, 80
"herring bone" method of tapping, 150
Heron, Fred, 174, 179–81, 191–94
Hevea Braziliensis (rubber trees), 148, 150, 162
"hierarchical-cum-imperial mindset", 53
"High Noon of Empire", 245
higher education, developments in, 203–10
H.M.S. *Eagle*, 85
H.M.S. *Sydney*, 98

H.M.S. *Terrible*, 72–73
H.M.S. *Warrior*, 65
Hobson, J.A., 14
Holt, Phillip, 79
Hong Kong
    British rule, 5
    entrepreneurs, 235
Hong Kong Shanghai Banking Corporation, 80, 232
Hoo Ah Kay, 229
Hooghly, gunboat, 49
Hooker, Joseph, 148
Hopkins, A.G., 15
Horsburgh, James, 45, 65
Horsburgh Lighthouse, 45–46, 60
Hotel de l'Europe, 112
housing, 10
Hovis bread, 177, 192–93
Howarth Erskine Limited, 44, 50, 204
Howe, Stephen, 24
Huff, Greg, 68
Hui Po-Keung, 39
hydraulic cement, 118
Hye Leong, ship, 79

I
ice cream, production of, 182
ice, introduction of, 181–82, 194
"idealistic engineering", 94
Illustrated London News, 36, 70
Imhoff, Karl, 127–28
imperial botanist, 143
Imperial College of Tropical Agriculture, 161
"imperial" commodities, 241
"Imperial Danger" thesis, 134
Imperial Economic Conference, 225
"imperial" food, 9
imperial ideas, and tropical medicine, 129–38

"imperial science", 64
imperialism, 5, 15–18, 21–22, 27, 43–44, 53, 140, 148, 184, 214, 240, 244
"imperialism of free trade", 221
imported technology, 6–7
import-substitution industries, 226
indentured labour, 19, 39, 130, 242
India, 3
  British rule over, 26–29
  convicts from, 136
Indian coolies, 133
  see also Chinese coolies
Indian Mutiny of 1857, 25, 94
Indiana, gunboat, 2
"indirect rule", concept of, 22
industrial entrepreneurs, 228
Industrial Revolution, 3, 13, 27, 30, 35, 42, 88, 145
"industrial science", 215
industrial waste, 120
infant mortality rate, 135
infectious diseases, 120–21
"informal control", 14
Inkster, Ian, 28, 229, 237
Innes, Emily, 106
Innes, J.J., 188
International Banking Corporation, 80
"Invincible Fortress", 10
Ipoh Tin Dredging, 37
irrigation system, in Egypt, 76

J
Jackson, John, 85
Jacques, Martin, 245
jamban, 112–13
Japanese colonialism, 243
"Japan Lights", 60
Japan, Meiji era, 20, 60
Jennings, Eric, 79
jinrickisha, 104–105, 109, 116

Jinrickisha Ordinance, 104
John Aird & Co., 76, 90
"John Chinaman", nickname, 40
John Little, store, 56
Johnston, Alexander, 35
Johnston's Pier, 111

K
Kang Yu Wei, 201
Katz Brother, 194
Kaur, Amarjit, 52
Kay, John, 13
Keasberry, Benjamin, 36, 57
Kedah, ship, 91
Kellogg, 169
Kennedy, Robert, 17
Keppel Harbour, 70, 72, 76, 88, 171
Keppel, Harry, 160
Kesner, Richard, 19, 32
Kew Gardens, 64, 143–44, 151
"Kew men", 148
Khoo Teng Soon, 188
Kim Ann, 189
Kimberly, Lord, 99
King Edward VII College of Medicine, 137, 210
Kinnersley Committee, 202
Kipling, Rudyard, 95
Klim, 179, 193
Koch, Robert, 120, 137, 142
ku-li (coolie), 58, 89–90
Kwok Acheong, 91
Kwong Yik Bank, 80
Kyd, Robert, 144

L
La Gloire, warship, 65
labour strike, 104–105
Lady Mary Wood, ship, 78, 90
Lagoon Dock, 76

Index

laissez faire policy, 8, 20, 123, 196, 197, 202, 232
Landes, David, 27
law of comparative advantage, 21
Lee, Bruce, 63
Lee Cheng Yam, 79
Lee Kuan Yew, 1, 197
Lemon, A.H., 205
Lemon Committee, 206
Lepers' Ordinance, 134
leprosy, 134–35, 141
Light, Francis, 230
"Lighthouse for All Nations", 46
"lighthouse Stevensons", 60
lighthouse technology, 45
Lim Boon Keng, 197, 200–201, 231, 242
Lim Eng Teck, 192
Lim Khoon Heng, 189
Linde British Refrigeration Company, 171
Lipton, 169
Little, Robert, 130
Liverpool School of Tropical Medicine, 129
Livingstone, David, 25, 33
Locke, P.V., 201
Lockhart, Bruce, 183–84
Lode Tin Mines, 37
Logan, Abraham, 36, 57
Logan's Journal, 39, 41, 154
London School of Tropical Medicine, 129
London Stock Exchange, 152
Low, James, 41
Lugard, Frederick, 16
Lyon, George, 70

M
Ma Huan, 157
Macfadyen, Eric, 106
Mackinnon, William, 78
MacRitchie, James, 123, 125, 127
MacRitchie Reservoir, 126
"Mad Ridley", 151
"Made in Singapore" policy, 177
Madras Artillery, 42
Madras Engineers, 42, 46
Magellan, Ferdinand, 65
Magnolia ice cream, 182
Mahan, A.T., 81
Majapahit, 3
"Malacca Connections, The", 79
Malacca, gunboat, 49
Malacca, ship, 79
malaria, 15–16, 57, 132–34, 139
Malay Express railway, 91
Malay Peninsula, 16, 22, 45, 53, 68, 103, 160, 174–75, 183, 203
Malaya
    British engineers in, 51
    railway line, 77
Malayan agriculture, 8
Malayan Civil Service, 52, 183
Malayan railways, 52
Malayan Tribune Company, 101
Malays, and education, 211–14, 219
managing-agency system, 36, 222
Manchester School of Radicalism, 83
Manderson, Lenore, 130
Mannesmann Tube Company Limited, 56
Mansfield & Company, 79
Manson, Patrick, 130
map-making, 64–65
Marco Polo, 157
marine engineering, *see* maritime technology
Marine Establishment, 49
maritime history, of Singapore, 79, 230
maritime technology, 65, 86
    advances in, 66–68

Marshall, Peter, 243
Matabele tribe, 16
Maternal and Child Health Services, 135
maternity hospital, 135
Mauritius, Botanic Gardens, 149
Maxim gun, 16
Maxwell, George, 211
McAlister & Company, 39
McKillop, John, 223
McLean Committee, 209
McLean, William, 209
McNair, Captain, 42
McVities, 169
Memphis, cruiser, 85
Mercantile Bank, 232
Mercantile Bank of India, 80
Mersey Docks & Harbour Board, 75
Metcalfe, Thomas, 112
Methodist Mission, 213
Metropolitan Board of Works, 94, 136
"metropolitan" interpretation, 14
middle-class colonists, 36
"Middle Kingdom", 2, 18
midwifery, 135
Midwives' Ordinance, 135
milk
  history of, 177–78
  production, 179–80
Milner, Alfred, 26, 213
Milwaukee, cruiser, 85
Mintz, Sidney, 153
mission schools, 202
missionaries, and imperialism, 24–25
modernism, 112
moneylenders (*chettiars*), 232
Monkhouse, Edward, 108
Montgomerie, W., 145, 147
Morris, Ian, 16
Morrison, A., 51
Morse telegraphic system, 100–101

mortality rate, 132–33
Mount Faber, 47
Mughal government, collapse of, 31
Muhlinghaus, Herman, 223
Municipal Act, 122
Municipal Authority of Singapore, 138
Municipal Building, 113
Municipal Health Department, 133
Municipal Ordinance, 121–22
municipal planning, 122
Murton, Henry James, 146, 148
musket, invention of, 30

N

Narayana, N., 189
naval base, 82–83, 240
Neave, David, 51
Nederlansch Indische Handelsbank, 80
Nemesis, steamship, 2
"neo-Europes", 20
Netherlands East Indies Agricultural Research Station, 179
Netherlands India, 74
"new arrivals", 38
"New Harbour", 7, 39, 45, 70, 72, 79, 104, 223
"new imperial history", 5
"new imperialism", 14
"new" quality food, 176–82
New World, 242
Newton, Howard, 123
night soil, 124–25
non-intervention policy, 96
Norddeutshcer Lloyd Company, 79

O

Ocean Steamship Company, 79
Oceanic Steam Navigation Company, 87

Index **275**

Orient Steam Navigation Company, 70
O'Grady, J.G., 53, 55
Old International Division of Labour (OIDL), 222
open-trading system, 21
Orchard Road Cold Storage, 172, 190
Ord, Henry, 21, 70, 198
Oriental Telephone & Electric Company (OTEC), 100–101
Ormsby-Gore, William, 161
Osborne House, 65

P
Pacey, Arnold, 94
paganism, 24
Pahang Consolidated Company Limited, 37
palm oil industry, 227
Pangkor Engagement, 22, 99
Parkins, John, 151
Pasteur, Louise, 120
Paterson Simons & Company, 40, 56, 73, 100
Pax Britannica, 29, 148, 169, 232
Paxton, Joseph, 29
Pearl's Hill Service Reservoir, 126
Pedra Branca, island, 45–46
Peek Freans, 169
Peirce, R., 123, 126
Peirce Reservoir, 126
Pell, Bennett, 100
Pender, John, 115
Peninsula and Orient Company, 72, 74–75
Peninsula & Orient, passenger ship, 36
Peninsular and Oriental (P&O) Steam Navigation Company, 78
pepper, cultivation of, 154–56
Perfection Stove Company, 187

"peripheral" thesis, 14
Perry, Mathew C., 72
Persse, Deburgh, 170–71
Philippines, colonization of, 242
Pineapple Experimental Station, 157
pineapple industry, 156–57
Pitt, William, 15
plantation agriculture, 144
plantation colonies, 19
Planter, The, 175
Platt, C.D.M., 15
P&O liners, 78
Population Census, 182
"population crisis", 161
population, Singapore, 93, 103, 123, 136, 182
pork, production of, 180
Porter, Bernard, 36
"Portland cement", 45, 111, 118
 see also "Condor" cement
post-Enlightenment attitudes, 53
powdered milk, 178
power generation, centralized, 109, 117
primitive society, 17
Procter & Gamble, 169
public health, 7, 120–38
Public Works Department (PWD), 52, 102
Pulau Brani, 223
Pulau Saigon, 111
Pulau Ubin, 46
Punch, magazine, 86
P.&W. MacLellan of Glasgow, 113

Q
Quaker Oats, 169
Quarantine and Prevention of Disease Ordinance, 177
Quarterly, 36
"Queen's Chinese", 200

Queen's Scholarships, 201, 206–208, 242
Queensland Meat Export and Agency Co. Ltd., 171, 191
quinine, 133

R
race, as scientific tool, 26
racial prejudice, 53–57, 62–63, 197, 245
racism, 5
Raffles College of Arts and Sciences, 208–209
Raffles Hotel, 48, 55, 112–13, 182
Raffles Institution, 197, 207
Raffles Museum, 148, 160
Raffles, Thomas Stamford, 1, 38, 41, 47, 65, 93, 145, 160, 163, 197, 230
   landed in Singapore, 2
"Railway Saga", 104
railways, in Egypt, 87
Raleigh bicycle, 103, 116
Ranee, gunboat, 49
Rapid Gravity Water Filter, development project, 56
Raub Australian Gold Mining Co. Ltd, 170
Reeve, J.W., 123
refrigerated transport technology, 175, 181, 190
Regenerator system, 100
Regnier, Philippe, 10
Reid, Anthony, 159
Renong Tin Dredging Company, 37
"Report on the Sanitary Condition of the Labouring Population of Great Britain", 136
Residential System, 99
Return of Industrial Scholarships and Apprenticeship, 204
Rhodes, Cecil, 16, 26

Rhodesia, 16
rickshaw coolies, 104, 131
Ridley, Henry Nicholas, 8, 147, 160
   rubber industry and, 147–52
Riley, Hargreaves & Company, 44, 49–50, 203–204
Riley, Richard, 49
Robinson & Company, 103
Robinsons, store, 56
Rock Drills, 106
Ronalds, Francis, 94
Roorkee Engineering College, 204
Roosevelt, President, 86
Rosecrance, Richard, 21
Ross, Ronald, 130
Royal Bakery, 176
Royal Botanical Gardens, 161
Royal Colonial Institute, 211
Royal Dockyards, 15
Royal Geographical Society, 45
Royal Society of Arts, 147
Rubber Estate Agency, 224
rubber industry, 56, 133, 152, 162, 174–75, 221–24, 227, 231, 237
   Henry Nicholas Ridley, and, 147–52
rubber plantation, living conditions in, 130, 133–34
Rubber Research Institute, 152, 158, 161–62
"Rubber Ridley", 151
rubber tapping, "herring bone" method, 150
rubber trees (*Hevea Braziliensis*), 148, 150, 162
Runciman, W., Reverend, 205

S
salted fish, 158
San Miguel Brewery, 182
sanitary engineer, 121–22, 126–27, 136

"sanitary science", 136
sanitation, and sewage improvements, 121–29, 136
Sappo, ship, 79
Schafer, Edward Albert Sharpey, 160
Schafer, Jack S., 160
School of Tropical Medicine, 139
"science for development" ideology, 161
"scientific" sanitation, 9
"Scramble for Africa", 18
Second Industrial Revolution, 227
Second World War, 188, 214, 217, 242
Sea Mew, steamer, 50
Sea View Hotel, 182, 184
Seabelle, yacht, 85
Seah Eu Chin, 41, 165
"Selangor Bakery", 176
Semmel, Bernard, 15
Senior Cambridge Examination, 201
Seven Years War, 15, 26
sewage, and sanitation improvements, 121–29, 136
"Shanghai jar", 112–13
Shannon, Margaret, 182
Shellabear, W.G., 214
shipbuilding, advances in, 49, 66, 71
shipping and commercial organizations, 77–81
Shipping Conferences, 225
"shipping revolution", 68
Shone hydro-pneumatic system, 125, 127
"short-termism", 233, 238
Sidney, R.J., 187
Simpson, W.J., 126–27
Singapore
   agency houses, 152
   banking businesses in, 80
   British economic interest in, 3
   British engineers in, 44, 46
   Chinese immigrants, 38–39
   commercial heartland, 79
   education, 9
   ethnic groups, 122
   European community in, 182–84, 187
   European firms in, 36
   fortification of, 42
   imperial port, as, 68–77
   infant mortality rate, 135
   link with the world, 94–102
   location, 20
   maritime history, 79, 230
   mortality rate, 132–33
   motor cars in, 105
   naval base, 82–83, 240
   population, 93, 103, 123, 136, 182
   port, 169
   road conditions in, 102–103
   shipbuilding in, 49
   social circles in colonial period, 56
   strategic importance of, 81–83, 99, 163, 243
   technological change in, 4
   trade revenue, 69
   trading emporium, as, 11, 221
   transportation technology, 102–107
   urban growth, 44, 128
   "walking city", 102, 121
   waterworks, 126
   wooden structures in, 110
Singapore and Malayan Directory, 56, 222–23
Singapore and Straits Aerated Water Company, 51
Singapore and Straits Printing Office, 51
Singapore Botanic Gardens, *see* Botanic Gardens of Singapore
Singapore Chamber of Commerce, 38, 224

Singapore Cold Storage, 9, 191,
    176–78, 183–84, 187–93
  advertisements, 185–86
  dairy farm, 179–80
  ice production, 181–82, 194
  imperial symbol, as, 187–90
  incorporating the, 170–75
  profits, 192
Singapore Dairy Farm Ltd., 193
Singapore Free Press, 41, 57, 69, 109,
    126, 128, 131, 137, 153, 155,
    201
Singapore Gas Company, 107
Singapore, gunboat, 49
Singapore Harbour Board, 55,
    76–77, 223
Singapore Institution, 197
Singapore Municipality, 55
Singapore Naval Base, 7, 85–86
Singapore River, 7, 46, 49, 68, 70,
    88, 110, 113, 128, 145
Singapore Rural Board, 55
"Singapore Story, The", 4
Singapore Tramway Company, 104
sinkehs, 38
Sino-Western enterprise, 40
siphon recorder, 97
Skerryvore, 46
Smith, Cecil Clementi, 149, 201
"Smith's Patent Germ Bread", 193
social circles, in colonial Singapore,
    56
social enclaves, 53
social exclusivism, 55
Song Ong Siang, 197, 201, 242
Soriano, Senor, 182
spice trade, 2, 145
St Andrew's Cathedral, 113
St James Power Station, 109, 113
St Joseph Institution, 197
Standard Oil Company of New York,
    187

steam transport, 8, 22, 39, 66–67, 71
Stearns, Peter, 3
Stereophagus Pumps, 128
Stevens, H.W.H., 171, 174, 191
Stevens, K.A., 192
Stevensons, David, 60
Stevensons, Thomas, 60
Stirling Castle, steamer, 72
Straits-born Chinese, 200–201, 210,
    214, 227, 231
Straits Chinese British Association,
    205
Straits Chinese Magazine, 200–201
Straits Civil Service, 241
Straits Ice Company, 194
Straits of Malacca, 2, 45, 64, 103
Straits of Singapore, 45, 60, 64
Straits Settlements, 22–23, 32, 44,
    47, 50, 54, 72, 75, 96–99, 102,
    122, 125, 130, 134, 139, 146,
    151–52, 161, 199–200, 203,
    205, 208, 213, 227, 231
  professionals in, 206
  Transfer of 1867, 136, 141, 198
Straits Settlements Port Ordinance,
    76
Straits Steamship Navigation
    Company, 79–80, 230
Straits Times, 48, 67, 73, 78–79, 89,
    110, 124, 128, 130, 132–33,
    170–73, 177, 184–86, 191, 198,
    226
"Straits Tin", 223
Straits Trading Company, 68, 223
strowger switching system, 101
submarine cables, 96–98, 147
Suez Canal, 7, 45, 67–68, 71, 87–88,
    93, 132
sugar production, 153–54
Supreme Court, 113
Swettenham, Frank, 50–55, 75, 199,
    211–12

Index

Sword, James, 223
Syers, H.C., 176
"systematic colonization", 43

T

Taiping Museum, 150
Taiping Rebellion, 38
Tamil labour, 147
Tamil-medium schools, 210–11
Tan Jiak Kim, 79
Tan Kah Kee, 228, 231, 237
Tan Keong Saik, 79
Tan Kim Seng, 229
Tan Kim Tian, 40
Tan Lark Sye, 231
Tan Tock Seng Hospital, 130, 132
Tanjong Pagar Dock Board, 76
Tanjong Pagar Dock Company, 44, 50, 70–72, 74–76, 79, 203
taxation, 21
tea consumption, 145
tea plantation, 73
Tea Research Institute, 162
technical arts, 43
technical education, 203–205, 216, 233
  see also vocational training
Technical Education Committee, 206
technical schools, 215
technology transfer, 3, 6, 9–10, 43, 52, 123, 221, 237, 243
  definition, 12
  limited industrialization, and, 222–26
  processes, 17
  racial prejudice, and, 53–57
  Western technology, diffusion of, 226–31
telegraph, invention of, 94–95
telegraphic systems, 100–101
telegraphy, 7, 17, 22, 94–101, 240

telephone system, 100
teleprinter system, 101
Telok Ayer reclamation, 88, 110
Temasek, 2
Temoh Tin Dredging, 37
Temperton, William, 49
Teo Lee, 229
Teutonia Club, 48
Thomas, Shenton, 85
Thomson, John, 36
Thomson, John Turnbull, 44–46, 60, 68, 125
Thomson Road Reservoir, 126
Thomson, William, 97
"thunderbox", 113
Times, The, 13, 26, 85
tin mining industry, 37, 175, 222–24, 227, 231
  machinery, 235
Tirpitz, Admiral, 82
Titanic, ship, 86
Tivendale & Company, 69
Tivendale, Thomas, 49
Tomlinson, S., 123
Topham, Jones & Railton, 76
trade depression, 156
trade entrepôt, 2, 7
trade revenue, Singapore, 69
trade schools, 207
tramway service, 104
transportation technology, 102–107
  refrigeration, and, 175, 181, 190
Trenton, cruiser, 85
tripartite partnership, 234
Trocki, C.A., 40
tropical colonies, 20–23
tropical diseases, 15, 114, 132
tropical medicine, and imperial ideas, 129–38
tropical production of goods, 3
tuan besar, 104
Tudor, Frederic, 181

Turnbull, Mary, 245
Twells, Alison, 24

U
underground cable system, 101
United Engineers Limited, 50, 109, 223
United States, Chinese immigrants in, 62, 235
urban growth, Singapore, 44, 128
urbanization, 7–8
USS *Mississippi*, 70, 72

V
Valdura, ship, 77
van Linschoten, Jan, 65
van Spilbergen, Joris, 65
"vapour compression", 181
vernacular education, 211–14
Victoria Dock, 70–71
Victoria Memorial Hall, 1, 48, 113
Victoria, Queen, 1, 13, 29, 65, 95, 107, 200
Victoria River Downs, 171
Victorians, as leaders of civilization, 26
Virchow, Rudolf, 142
vocational training, 204–205
see also technical education

W
Wah Hin, 189
"walking city", 102, 121
Wallich, Nathaniel, 145, 160
Walton, Oliver C., 59
Wang Gungwu, 233
War Office, 99
Warren, James, 131
Warsop Road Breakers, 106
water closet, introduction of, 127, 139

water filtration, 127, 137
water quality, 120, 125–26
waterworks, 126
Watts, Sheldon, 25, 134
Waugh, Henry, 194
Wee Ah Hood, 229
Wee Bin & Co., 230
Weld, Frederick, 10, 53, 55, 72, 81, 122, 213
Wellesley, Arthur, 31–32
Western faith, proselytization of, 17
Western superiority, 24–25, 53, 211
see also British superiority
Western technology
 diffusion of, 226–31
 power of, 16–18
Whampoa, 108
Wheatstone telegraphic system, 100
"Whippet Roadster", 105
"white man's burden", 211
"White Rock" (Pedra Branca island), 45–46
white supremacy, 23, 26, 53
Whiteaway and Laidlaws, store, 56
Wickham, Henry, 148
Wilkinson, R.J., 205
Wilkinson, Tivendale and Company, 49
Will O' the Wisp, ship, 79
Williams, Michael, 14
Winstedt Committee, 207
Winstedt, Richard, 137, 206, 208
 education policy, 213–14
"Winter Garden Lounge", 188
Wong, Rosalind, 101
Wong, S.Q., 101
Woodall, Corbet, 108
Woods, Robin, 36
Woolley Committee, 198
"workshop of the world", 13–14, 123, 244

Index

World Expo, 29
World War I, *see* First World War
World War II, *see* Second World War
Wray, Leonard, 150
Wright, Henry, 134

Y
Yacht Club, 191

Yeo Kwan, 191
Yeo Swee Hee, 173, 191
Young, George, 26
Yuill, George Skelton, 191, 193

Z
Zhang Huachen, 86
Zimbabwe, 16

www.ingramcontent.com/pod-product-compliance
Lightning Source LLC
Chambersburg PA
CBHW072128290426
44111CB00012B/1826